A LEVER LONG ENOUGH

ROBERT McCAUGHEY

A LEVER LONG

ENOUGH

A History of Columbia's School
of Engineering and Applied
Science Since 1864

Columbia University Press / New York

Columbia University Press
Publishers Since 1893
New York   Chichester, West Sussex
cup.columbia.edu
Copyright © 2014 Columbia University Press
All rights reserved

Library of Congress Cataloging-in-Publication Data
McCaughey, Robert A.
    A lever long enough : a history of Columbia's School of Engineering and
Applied Science since 1864 / Rober McCaughey.
        pages cm
ISBN 978-0-231-16688-1 (cloth : alk. paper) — ISBN 978-0-231-53752-0 (ebook)
1. Fu Foundation School of Engineering and Applied Science—History. I. Title.
T171.F854M33 2014
620.0071'17471—dc23

                                                              2014007037

COVER DESIGN: Faceout Studio
COVER IMAGE: General Research Division, The New York Public Library,
Astor, Lenox, and Tilden Foundations

References to websites (URLs) were accurate at the time of writing. Neither
the author nor Columbia University Press is responsible for URLs that may
have expired or changed since the manuscript was prepared.

To Ann, as always,
to River and Bode, Wyatt and Fiona,
and to Columbia's engineers

*Give me a lever long enough and somewhere
to stand and I will move the world.*

—ARCHIMEDES (287–212 B.C.)[1]

CONTENTS

CONTENTS

ILLUSTRATIONS

FOREWORD

F ROM ITS BEGINNING as King's College in 1754, Columbia valued the academic disciplines of science, engineering, and mathematics. Indeed, some of Columbia's early alumni were pioneering entrepreneurs and engineers. John Stevens (class of 1768) developed steam engines that powered both the first steamships to navigate the open ocean and the first steam locomotive. As governor of New York, DeWitt Clinton (class of 1786) was the driving force behind the Erie Canal, which connected New York City to the burgeoning Midwest. Horatio Allen (class of 1823) was an early president of the American Society of Civil Engineers.

But it was only in 1863 that Thomas Egleston—then a mineralogist working at the Smithsonian Institution—suggested that Columbia create a separate school of metallurgy and mining. In less than a year, on November 15, 1864, the School of Mines of Columbia College opened its doors, with a student body of twenty and a faculty of three. The new school reflected in many ways the demands of its age, recognizing the need for raw materials to wage the Civil War and to supply the rapid industrialization of the United States.

Since then, Columbia Engineering has grown and expanded on every front. Today our school is home to more than 170 faculty members, 1,500 undergraduate students, 2,000 master's degree students, and 700 doctoral students. The school quickly moved from a simple focus on mining to include civil engineering, chemical engineering, and electrical engineering. We now have a robust set of nine departments. We've moved from being a bastion of the city's male Knickerbocker elite to a global institution, attracting students from around the world and an entering undergraduate class that is 44 percent women. The school's footprint has grown as well, as we've moved from Midtown to Morningside Heights and from one building to multiple campus locations. Our school's curriculum still reflects Columbia's unfailing commitment to liberal arts instruction, while pushing the frontiers of technology and applied science.

The Fu Foundation School of Engineering and Applied Science can boast of a rich history of pioneering education, research, and innovation. Professor Robert McCaughey's *A Lever Long Enough*, written on the occasion of the school's sesquicentennial, offers an insightful look at the birth and evolution of an engineering school that has grown within the context of a renowned research university, evolved with the disciplines that it pursues, and interacted dynamically with its city, its country, and the world. Like any great institution, Columbia Engineering has been shaped by strong personalities, weathered terrible storms, and enjoyed remarkable individual and institutional success. Columbia Engineering's steadfast commitment to excellence has manifested itself not simply in academic pursuits but also in pushing beyond conventional disciplinary frontiers.

* * *

Throughout its history, Columbia Engineering has always focused both on academic pursuits and on engagement with the world. The first dean of the Columbia School of Mines, Charles F. Chandler, set a high benchmark. A chemist by training, Chandler also emerged as a major public health advocate. Working with New York City's Metropolitan Board of Health from 1866 to 1883, he developed standards for clean and safe drinking water, milk for babies, and medicines. A vocal supporter of improved tenement housing, including mandatory indoor plumbing, he has been credited with preventing a cholera epidemic in 1883.

Chandler's thirty-two-year leadership of our school marked a period of significant engineering and scientific advancements, not only in chemistry and chemical engineering but also in electrical engineering, civil engineering, and nascent computer science.

When Herman Hollerith (class of 1879), who had been a statistician for the 1880 U.S. Census, returned to Columbia for graduate studies, he set about finding a new way to process census data. His doctoral thesis described a punch card system in which the cards could be counted and sorted mechanically. Awarded a PhD in 1890, Hollerith went on to found the Tabulating Machine Company in 1896, which merged with three other companies to form what became International Business Machines (IBM) in 1924. Today Hollerith is recognized as the father of modern automatic computation.

Few engineers, however, can match the achievements of Michael Pupin (class of 1883). By 1882, when Thomas Edison wrote to Columbia President Frederick Barnard about the need to educate "electrical scientists" to work for his new company, Pupin had been among the Columbia students already experimenting with electricity. After graduate study in England and Germany, in 1889 Pupin joined the faculty of the newly formed Department of Electrical Engineering. Over the next four decades, Pupin won more than thirty patents in fields ranging from electricity to medicine, from telephony to sonar. Using a fluorescent screen given to him by Thomas Edison, Pupin developed an X-ray method that produced usable images while reducing radiation exposure from one hour to just a few minutes. In 1899, he patented the "Pupin Coil," which made long-distance telephony feasible; he later sold the patent to American Telephone and Telegraph.

One of Pupin's pupils, Edwin H. Armstrong (class of 1913), invented the regenerative circuit while he was still an undergraduate. Upon graduation, he joined the school's faculty. Commissioned as an officer in the Army Signal Corps during World War I, Armstrong developed the superheterodyne circuit. After the war, he returned to Columbia, where he developed the superregenerative circuit and, in 1933, completed his most famous invention: FM radio. Although best known as the "father of FM radio," Armstrong's inventions actually underlie all modern radio, radar, and television.

In civil engineering, William Barclay Parsons (class of 1882) was building some of the major transportation projects of the day—railroads, rapid transit, and canals. Parsons and his younger brother Harry, a mechanical engineer, opened up their own consulting engineering office in New York in 1885. The Parsons firm had a global impact on the built environment, with projects ranging from hydroelectric plants across the United States to docks in Cuba, from China's Hankow-to-Canton railroad to the Cape Cod Canal.

In 1894, Parsons became chief engineer of the New York Rapid Transit Commission, designing the Interborough Rapid Transit subway, which opened in 1904. That subway network secured Parsons's place in history and revolutionized transportation in the metropolis. Parsons maintained a life-long relationship with Columbia, serving on the Board of Trustees from 1897 to 1932, the last fifteen years as its chair. His firm, now called Parsons Brinckerhoff, remains a leader in the design, construction, and operation of critical infrastructure around the world.

A pioneering chemist, Irving Langmuir (class of 1903) worked at the intersection of chemistry, physics, and engineering, investigating the properties of adsorbed films and the nature of electric discharges in high vacuum and in certain gases. He was awarded the Nobel Prize in chemistry in 1932 "for his outstanding discoveries and investigations within the field of surface chemistry."

After graduate study at Leipzig, Colin Fink (class of 1903) worked for General Electric and later as chief chemist for the Chile Exploration Company. In 1921, he joined the faculty, heading up a new branch of chemical engineering, electrochemistry. Fink developed a chromium-plating process that is still used today as the standard for plating metal or plastic. His pioneering electrochemistry research led to ductile tungsten for incandescent lamp filaments, an insoluble anode for electrowinning copper, an electrolytic process to remove corrosion from antique bronzes, and several electroplating processes.

By the 1930s, Columbia Engineering alumni and professors were pushing the boundaries of their disciplines. Raymond D. Mindlin (class of 1931) , who joined the Civil Engineering faculty in 1940, was the most outstanding elastician of his generation, making contributions to the fields of applied physics, applied mechanics, and engineering science. In 1946, President Harry Truman awarded Mindlin the Medal for Merit for his work developing the radio proximity fuse, a detonator used extensively in World War II. In 1979, President Jimmy Carter awarded him the National Medal of Science. Maurice A. Biot, who joined the Civil Engineering department in 1937, was another pioneering researcher, laying the foundations of the theory of poroelasticity (now known as Biot Theory), which describes the mechanical behavior of fluid-saturated porous media.

* * *

As the nation began to recover from World War II, the school began to reassert itself in a variety of areas. In electrical engineering, faculty and alumni set a new standard for industrial impact. The "father of robotics," Joseph Engelberger (BS 1946, MS 1949) developed the Unimate robot— the first industrial robot in the United States. His industrial robots, which first appeared on automobile assembly lines in 1961, revolutionized countless manufacturing processes. Today his newest robots assist in human care, especially with the elderly.

John R. Ragazzini (PhD 1941), who had worked on the Manhattan Project, collaborated with Loebe Julie to develop and build the world's first modern differential operational amplifier in 1947. In 1952, Ragazzini and faculty colleague Lotfi A. Zadeh (PhD 1949) established the z-transform method that is still the standard in digital signal processing and other discrete-time systems. Rudolf E. Kálmán (PhD 1957) developed the Kalman Filter, a mathematical algorithm widely used in many prediction models. In 2009, President Barack Obama awarded him the National Medal of Science. Robert Moog (MS 1956) revolutionized music with his invention of the Moog synthesizer, the first voltage-controlled subtractive synthesizer played via keyboard.

In chemical engineering, Professor Elmer Gaden (BS 1944, MS 1947, PhD 1949) demonstrated that an optimal amount of oxygen enabled penicillin mold to grow and multiply more rapidly. This research formed the basis for mass production of a wide range of antibiotics, beginning with penicillin, and earned Gaden the title of "father of biochemical engineering."

In mechanical engineering, Professor Ferdinand Freudenstein (PhD 1954) developed the precursor of what is now known as the Freudenstein Equation, which uses a simple algebraic method to determine the position of an output lever in a linkage mechanism. Freudenstein later applied digital computation to the kinematic synthesis of mechanisms, earning him recognition as the "father of modern kinematics."

Near the end of World War II, IBM established the Watson Scientific Computing Lab at Columbia, and the lab staff started teaching the first ever computer science courses fully integrated into a university curriculum. Joseph Traub (MS 1955, PhD 1959) gained early access to computers through graduate work in the Watson Lab. Traub's doctoral thesis examined computational quantum mechanics, and he continued his pioneering work in optimal iteration theory, developing significant new algorithms at Bell Labs, the University of Washington, and Carnegie Mellon University before returning to Columbia in 1979 as the founding chair of the Department of Computer Science.

The 1950s and 1960s witnessed interdisciplinary initiatives that applied engineering principles to medicine and the study of the human body. In 1968, Edward F. Leonard conducted path-breaking research in the engineering and design of artificial organs. Richard Skalak (BS 1943, PhD 1954) integrated engineering mechanics and biomedical science to understand the mechanics of blood flow, bone growth, white blood cell response to infections, and biological implications of and responses to implants.

Sebastian Littauer (MS 1928), who joined the Department of Industrial Engineering in 1947, introduced the first courses in operations research at Columbia in 1952. By the 1990s, with the resurgence of Wall Street, the renamed Department of Industrial Engineering and Operations Research had been expanding into financial engineering.

Van C. Mow joined Columbia in 1985 as the first joint faculty appointment between Columbia Engineering and the College of Physicians and Surgeons. A decade later, Mow and Leonard, joined by W. Michael Lai and Gerard Ateshian of the Department of Mechanical Engineering, became the founding faculty for a biomedical engineering program. Under Mow's leadership, the new program won additional support, hired new faculty, and expanded its partnership with P&S. In 2000, university trustees approved the Department of Biomedical Engineering, creating the school's ninth department.

Meanwhile, the school's mining and metallurgy department (christened the Henry Krumb School of Mines, in a nod to the school's original mission) pivoted to embrace environmental study, later becoming the Department of Earth and Environmental Engineering. It began to support new research initiatives, including the Earth Engineering Center and the Waste to Energy Research and Technology Council, and organized the First International Conference on Environmental Issues and Waste Management in Energy and Minerals Production.

As the twentieth century came to a close, a transformational donation propelled the school forward in its mission. A $26 million gift from businessman and philanthropist Z.Y. Fu in 1997 targeted four areas—computer science, biomedical engineering, applied mathematics, and electrical engineering—and gave the school a new name. The new resources enabled the school to expand, attracting and retaining new faculty talent and increasing interdisciplinary collaboration within the school and with other divisions of the university.

* * *

By the beginning of the twenty-first century, Columbia Engineering had developed a new paradigm of highly interdisciplinary engineering and applied science research. Columbia Engineering professors eagerly responded to major challenges that were inherently interdisciplinary and could only be met with an expansive approach.

In the early 1990s, Dimitris Anastassiou and his student Fermi Wang (PhD 1991) developed a key patent essential to the MPEG-2 video and systems coding standards used in TVs, DVD players and recorders, personal computers, computer gaming, and cameras. But then, Anastassiou shifted his research to systems biology, specifically cancer research. He has developed a new computational model to predict breast cancer survival and is searching for precise genetic signatures present in multiple types of cancer.

Jingyue Ju, who joined the Department of Chemical Engineering in 1999, has directed his research toward biomolecular applications for personalized medicine. His revolutionary research developed a new DNA-sequencing platform that heralds a new era in the field. His innovations could reduce the cost of DNA sequencing, so that it can become a routine part of medical research and individualized health care.

In 2005, Gordana Vunjak-Novakovic joined the Department of Biomedical Engineering. Expanding the department's strength in tissue engineering, she has developed novel bioreactors to advance her research on engineering functional tissues. She has successfully grown bone grafts that can match a patient's original jaw bone for facial reconstruction surgery and has engineered functional cardiac tissue.

Founded in 1978, what is now the Department of Applied Physics and Applied Mathematics quickly expanded beyond its original core of plasma physics and nuclear science to include areas such as solid-state physics. Today Adam Sobel uses sophisticated climate modeling in his research of extreme weather events, such as hurricanes, and their relation to climate, seeking engineering solutions to ameliorate the social impact of extreme weather.

Computer science has expanded into new areas as well. In 2001, Kathleen McKeown, a leading scholar in the field of natural language processing, developed Newsblaster, a trail-blazing system that automatically identifies, sorts, and summarizes the day's top news stories; now she is beginning research on using data from social media to track and respond to disasters. Shih-Fu Chang has developed information analysis and machine-learning techniques to process voluminous online video images into useful data and conduct unprecedented searches using large visual databases. Eitan Grinspun integrates concepts from graphics, applied mathematics, and engineering, applying the budding field of discrete differential geometry to create practical computations that enable film studios to produce detailed, lifelike animation.

In mechanical engineering, Vijay Modi uses engineering to transform both developing and highly developed regions of the world. His research optimizes electric grids, not only in New York City but also in self con tained micro-grids that supply local power in the developing world. He currently leads the UN Millennium Villages Project on the role of energy and energy services.

Columbia Engineering now stands at the nexus of many of the university's interdisciplinary initiatives. Engineering professors are making key contributions to the Earth Institute's initiatives in water, climate, energy, and sustainability. Engineering professors are helping shape the new Mortimer Zuckerman Mind Brain Behavior Institute, which is closing the gap between our fundamental understanding of the brain and what the social sciences, humanities, and professional disciplines tell us about behavior.

In 2012, the university announced the launch of the new Institute for Data Sciences and Engineering, which is poised to become a world-leading institution in research and education in the theory and practice of the emerging field of data science. Based at Columbia Engineering and led by engineering professors, the institute's six centers encompass faculty from nine university schools, an acknowledgment not only of the institute's inherently interdisciplinary subject matter but also of engineering's pivotal place in meeting all aspects of the challenge.

* * *

This year's sesquicentennial celebration is an unprecedented opportunity to contemplate our school's bright future even as we celebrate our storied past. I firmly believe that we are in a Renaissance for engineering—a period of great research, great innovation, great invention, and incredible translation of these innovations to solutions that will benefit people around the world. We are privileged to build on our school's remarkable legacy in a time when engineering and the applied sciences have never been more important to the progress of society.

As we embark on our next 150 years, our school continues its proud tradition of educating leaders, whose vision, creativity, and innovation have enabled us to enjoy a more advanced and livable world. Columbia Engineering's alumni are inventors and entrepreneurs, Fortune 500 CEOs and finance leaders, outstanding academicians (including multiple Nobel Prize winners), and even astronauts. These women and men—whether in academia, government, or industry—share an innate intellectual curiosity and drive that have always been hallmarks of a Columbia education. As Professor McCaughey's fine history of our school shows, Columbia Engineering has always provided a lever long enough to move the world.

Mary Cunningham Boyce, Dean of Engineering, Morris A. and
Alma Schapiro, Professor, New York City, March 2014

PREFACE

T HE RESEARCH FOR this history of Columbia University's School of Engineering and Applied Science (SEAS) began with a comparative question: How, if at all, does the history of Columbia's engineering school differ from those of other American engineering schools, specifically the other Ivies: Harvard, Yale, Brown, Princeton, Penn, Dartmouth, Cornell, and the currently top-ranked MIT and Stanford? All these institutions are privately governed, and all started engineering programs between 1847 and 1890, most in the two decades bracketing 1860. They all number among the nation's top-ranked universities. But they also differ in crucial respects. The engineering programs at MIT and Stanford, as well as Cornell's, have histories coterminus with their institutions. MIT was specifically founded to advance the study of technology, and at Cornell and Stanford technology enjoyed a pride of place among other instructional fields. All three have engineering programs substantially larger than Columbia's and the other Ivies', both in absolute size and relative to their universities. As such, they share a technology-centered character not unlike the first land-grant universities, the Universities of Michigan, Illinois, and California, and state-funded institutions like Georgia Institute of Technology, all with well-regarded engineering schools, and all founded in the second half of the nineteenth century.

What Rosalind Williams has written of MIT applies in some degree to them all: "I cannot imagine a more pro-technology place."[1]

At Columbia and the other Ivies, excepting Cornell, engineering appeared on the scene a century into their histories. These colleges had developed institutional priorities that made engineering a problematic addition to their original mission, and a modest, questionable, and in some instances, discontinuous presence. Thus, while what follows is an account of the history of a single engineering school and not comparative history of Ivy engineers, much less that of all engineering schools, I believe the history of Columbia engineering sheds light on other "technology-wary" academic institutions, on the history of engineering education, and on the changing place of technology in modern and post-modern America.

By identifying an institution as "technology-wary," I am not suggesting it is also "anti-science," although it could be. A skepticism about technology as a legitimate subject of study can exist—and has—in institutional circumstances even where science has come to be highly valued. Such skepticism has not been limited to trustees, administrators, alumni, or humanities faculty, but can also be found among the science faculty. Achieving the favored place of science in the university has at times required that scientists posit a distinction between the disinterested ideological purity of their own calling and the market-driven, problem-solving, inventive ways of the engineers and applied scientists.

The effect of this division at technology-wary institutions is not so much a hierarchy of "two cultures," as C. P. Snow described the situation in 1959, but of three: the humanists, the scientists, and the engineers or applied scientists. In this three-way situation, the humanists and the scientists have held the ideological high ground, while engineers were consigned to a lower place in the institutional ordering.[2] The history of Columbia's school of engineering both conforms and departs from this pattern. Given the decidedly humanities-centered character of Columbia College, the engineering school's opening two decades were impressively dynamic and transformational. It was in the School of Mines where instruction in geology, physics, astronomy, and biology at Columbia took permanent root. Longtime Mines dean and chemist Charles Frederick Chandler was Columbia's most noted scientist of the last quarter of the century, while Mines professor and physicist Michael I. Pupin was one of the two most visible Columbia scientists, with the geneticist Thomas Hunt Morgan, of the first quarter of the twentieth century. Other Mines faculty figured prominently in Columbia affairs and the scholarly activities of their fields, while Mines students numbered among Columbia's most successful graduates, including titans of industry, inventors, and scholars. But by the 1890s the school, no longer the School of Mines but variously labeled over the next seventy years, had lost some of its earlier standing within the university, and ceased to attract the active support of the university administration or the approbation of Columbians more generally. Once focused in instruction in the classics, literary and classical studies, turn-of-the-century Columbia came to be best known as a center for humanistic studies, social sciences, and increasingly of what it labeled and privileged as "pure science."

Engineering retained a place in the university's offerings in the new century, its faculty continued to play significant roles in their disciplines, and its students continued to put their training to productive and gainful use. But that place was diminished from what it had once been, and would remain a modest one into the 1970s. It was during that decade—the most trying in all of Columbia's history, and the decade in which both Harvard and Yale, along with nearby NYU, all but abandoned their engineering programs—that Columbia engineering effected a turnabout to emerge from it strengthened and better positioned to meet future challenges. Since then, though unevenly, the engineering school has been on an upward ascent. As it prepares to celebrate its one hundred fiftieth anniversary, Columbia's standing among engineering schools is rising faster than at any time in its history, its place in the university never more secure, and never more central.

With this enhanced standing comes the opportunity and the responsibility to compete with other great engineering programs for the very best faculty, the most promising students, and for the funding needed to advance knowledge to "make straight the way." But, so, too, with the recentering of the school comes an opportunity and an educative responsibility peculiar to Columbia's engineers and intrinsic to their institutional setting: to provide guidance not only to their own students but to their non engineering colleagues and those colleagues' students who, ready or not, must find their way in this dawning age of technology. As never before, all Columbians look to our engineers to provide us with that "lever long enough" that Archimedes called for more than two millennia ago, by which we can all "move the world."

Robert A. McCaughey
New York City
September 2013

ACKNOWLEDGMENTS

T HANKS TO DEANS Feniosky Peña-Mora and Mary Cunningham Boyce, Feni for the idea of doing this book and Mary for support in seeing it through to publication. And to Margaret Kelly, SEAS director of communications, for her many assistances and unwavering encouragement.

I also acknowledge the help provided by Michael Ryan, director, and his staff at the Rare Book and Manuscript Library, including Susan G. Hampson, university archivist, and especially Jocelyn Wilk, associate archivist, without whose help anyone venturing into Columbia's history would be lost. Undergraduate assistants to whom I am grateful include Noah Whitehead (SEAS 2013), Hannah Rubashkin (BC 2013), Lauren Gorab (BC 2014), and Nina Deoras (BC 2016).

Help and encouragement were also provided by the staff of Columbia University Press, particularly my editor, Philip Leventhal, and Whitney Johnson and Sapna Rastogi of Cenveo Publisher Services.

I wish to thank all the faculty, alumni, and administrators of SEAS, as well as Columbia officials, who allowed me to interview them in the course of my research and to cite them in my notes. I especially wish to acknowledge the help in this regard of Morton Friedman, Daniel Beshers, and Salvatore Stolfo, who read successive drafts and challenged me at several interpretive junctures along the way.

I thank my Barnard College history colleagues, Herb Sloan, Nancy Woloch, Lisa Tiersten, Jose Moya, Joel Kaye, and especially Deborah Coen, who helped me with the relevant secondary literature in her field, the history of science and technology.

On the home front, thanks to Barbara Butterworth and Michael Gill, for your friendship and for tolerating my absences from Wednesday evening races and boatyard chores.

Last and first I acknowledge the help and support of my wife, Ann, who encouraged me to take on this project, who served as its first reader, and who for the past three years welcomed Columbia's engineers into the warp and woof of our conversational lives together.

A LEVER LONG ENOUGH

1

ENGINEERING IN AMERICA — BEFORE ENGINEERS

The attention of several Deities appeared to be arrested by the coming of Neptune, who . . . was pointing to a scene of Minerva, who, having laid aside her aegis, supported with the left hand a Medallion of Clinton, while, with the right, she pointed to a distant view of Columbia College.

— OFFICIAL DESCRIPTION OF THE NEW YORK CITY CELEBRATION ON COMPLETION OF THE ERIE CANAL, PRESIDED OVER BY GOVERNOR DEWITT CLINTON (COLUMBIA COLLEGE, 1786), NOVEMBER 4, 1825[1]

[Columbia College] good in classics, weak in sciences, with few graduates of distinction.

— [BOSTON-BASED] *THE CHRISTIAN EXAMINER*, 1854[2]

WHAT FOLLOWS IS a history of a single American engineering school, Columbia University's Fu Foundation School of Engineering and Applied Science, as the University's formal name proudly proclaims, "in the City of New York." It is intended to be celebratory, published on the occasion of the school's one hundred fiftieth anniversary, and interpretive, as befits a school that has known over fifteen decades of both success and adversity. But neither institutional loyalty nor civic pride allows the history of American engineering to begin with Columbia in the 1860s or have its origins in New York City. For that larger story we must go back earlier and farther afield, before the founding of engineering schools or the formation of the engineering profession, to individual Americans, "engineers before the name," and the purposeful changes they wrought to the world they inhabited.[3]

CHANGES IN THE LAND (AND WATERS)

As William Cronon instructed us thirty years ago in *Changes in the Land: Indians, Colonists, and the Ecology of New England*, the first engineering challenges of North America's European settlers involved altering the natural landscape earlier inhabitants had fashioned to their purposes. This involved clearing the land of trees and rocks, enclosing fields and pastures with walls and fences, and building houses and barns, churches, markets and public edifices, and roads. More recently, W. Jeffrey Bolster, in *The Mortal Sea*, has reminded us that these first European Americans were coastal people, building boats, docks, breakwaters, bridges, and lighthouses. They ventured early into mining, manufacturing, and iron making and the fashioning of farm tools, cooking utensils, barrels, and guns. By the 1740s colonial Americans were distilling rum and manufacturing candles for export.[4]

To be sure, early Americans relied heavily upon Europe for finished goods such as clocks, fine furniture, maps and charts, and surveying instruments. They also depended upon European visitors or new arrivals for needed expertise in architecture, cartography, and military engineering. Emigrés like the Scottish-born surveyor general of colonial New York, Cadwallader Colden (1688–1776), the English-born cartographer of the New England coast, Cyprian Southack (1662–1745), the English-born importer of textile technology, Samuel Slater (1768–1836), and the British-born architect of the nation's capital, Benjamin Henry Latrobe (1764–1820), attest to early America's reliance on technology transfers effected by immigrants.[5]

The exigencies of life in early America brought to the fore leaders whom necessity made engineers before the name: William Bradford, John Winthrop, and Roger Williams, the founders of Plymouth (1620), Massachusetts Bay (1630), and Rhode Island (1635), respectively, all engaged in boat-building and water-diversion projects for which their lives in England had not prepared them. Meanwhile, their neighbors took up surveying, blacksmithing, ship's carpentry, barrel-making, and printing. It is with one of these first printers that the story of American technology might be said to begin.[6]

The historian Joyce Chaplin's recent biography of the eighteenth-century Boston-born and Philadelphian-by-adoption Benjamin Franklin (1706–1789) aptly calls him the "first scientific American." He could be equally viewed as the "first American engineer," occupationally engaged in applying scientific and mathematical principles to practical challenges. Franklin's 1751 invention of the lightning rod, made possible by his partial understanding of the character of static electricity, following hard on experiments confirming lightning to be a form of electricity, along with his design of the "Franklin stove," using his understanding of heat convection, are instances of applied science, of engineering. So, too, his effort to improve the efficiency and safety of sailing ships by reconfiguring their hulls, sails, and rudders constitutes early exercises in marine engineering.[7]

Three of Franklin's younger contemporaries—John Fitch (1743–1798), John Stevens (1749–1838), and Robert Fulton (1765–1815)—deserve inclusion in the ranks of early American engineers. Fitch, Connecticut-reared and self-taught, was the first to demonstrate the potential of a steamboat, on the Delaware River in August 1787, for no less a distinguished audience than the members of the federal convention gathered that summer in Philadelphia. Stevens, from Hoboken, New Jersey, demonstrated the steamboat's seaworthiness in the open ocean on a 1798 run from Cape May to Hoboken. Fulton, an English-born polymath, demonstrated the commercial possibilities of the steamboat in 1807 with a regularly scheduled run of the paddlewheeled *Clermont* on the Hudson River between New York and Albany. Robert R. Livingston, Jr., while better known as a Revolutionary era diplomat and a leading New York Jeffersonian, financed Fulton's experiments and secured exclusive rights to steamboating on the Hudson, then introduced steamboats on the Mississippi. Livingston (1765), like Stevens (1768), was a graduate of King's College. These men provide two biographical links between the early story of American engineering and King's College, later Columbia University.[8]

Other early American engineering projects involved the utilization of water as transportation medium and source of mechanical power and the conveyance of drinking water to urban populations. The prospect of canals linking the American coast to the interior was recognized by Americans several decades before it was realized. In 1724 New Yorker Cadwallader Colden proposed linking the Great Lakes to the Atlantic by canals extending from the Mohawk River westward through an opening in the Appalachian Mountains to link Lake Erie to the Hudson River. It took more than ninety years before political will and technical wherewithal became available. Slightly more modest were the plans of a Virginia planter, George Washington, surveyor and land developer before soldier and statesman, formulated in the 1760s and undertaken two decades later in the 1780s, to link Chesapeake Bay to the Ohio River. The topographic challenge of this proved, in Washington's time, insurmountable. But even as this Founding Father labored on his immense canal project, his younger Virginia compatriot, Thomas Jefferson, the son of a surveyor-cartographer and planter, worked on linking his Monticello plantation to the James River (and to the sea) by straightening and redirecting the flow of the Rivanna River, which meandered through his property. Both made better presidents than engineers. Not to be outdone by his fellow founders, Benjamin Franklin sketched a plan by which his Pennsylvanians might build a water thoroughfare to the west by way of canals linking the Delaware and Schuylkill to the rivers of western Pennsylvania.[9]

Elsewhere, canal projects succeeded. One was undertaken in the 1790s in Charleston harbor to link South Carolina's Santee and Cooper Rivers, and another in northeastern Massachusetts, where the 22-mile Middlesex Canal linked Boston to the Upper Merrimack River and the planned industrial town of Lowell, the future center of textile manufacturing. Construction of the Middlesex Canal was overseen by the self-taught Loammi Baldwin, "the father of American civil engineering." Both canals made their private investors a good return and were hailed as a public good.[10]

The early nineteenth century's most ambitious and successful canal engineering project was launched in 1817 by the incoming governor of New York, DeWitt Clinton (1770–1827). Whereas Colden's plan envisioned a 250-mile canal running west from the end of the Mohawk River and north to Lake Ontario, Clinton promoted a plan for a 375-mile canal running due west to the eastern shore of Lake Erie. It is here, during the eight-year construction of what critics called "Clinton's ditch," that American engineering had its beginnings as a collective enterprise, the first instance of combining technological expertise and organizational skill to produce a beneficial modification of the environment. Those who brought these skills to the building of the Eric Canal—Benjamin Wright, Canvass White, James Geddes, John Jervis—were self-taught, making their success all the more remarkable. The politician behind the project, DeWitt Clinton, had been a member of the first graduating class of the newly named Columbia College in 1786.[11]

Another use to which American waters were put followed hard on the removal of restrictions on manufacturing that came with independence. In 1793 the English immigrant Samuel Slater (1766–1836) set up a cotton textile mill, using plans he had acquired in the employ of Richard Arkwright, at the Pawtucket falls of the Blackstone River. Soon thereafter milldams appeared up and down the rivers of southern New England and New Jersey, including one on the Passaic River in which Alexander Hamilton (King's College 1774–1775) had a financial stake, which would provide the hydropower to propel the first American industrial revolution.[12]

Shipbuilding became another engineering undertaking of the early republic. From President John Adams's authorizing the construction of six naval frigates in 1794, which engaged the talents of Joshua Humphries, to the design and construction of steamboats in the early 1800s by Fitch, Fulton, and Stevens, to the experiments of John Ericsson with iron hulls and screw propellers in the 1840s, to the design and construction by Donald McKay and William H. Webb of the clipper ships of the 1850s, Americans in the decades before the Civil War numbered among the world's most accomplished marine architects/engineers.[13]

That Fulton, Ericsson, Webb, and McKay for a time had their shipbuilding operations in New York attests not only to the city's standing by the 1820s as America's leading port but also to its importance as a crucible for engineering enterprise generally. (Stevens operated out of nearby Hoboken.) A further instance of a water-related undertaking of epic proportions—the Croton Aqueduct project—reinforces that conclusion. By the 1830s New York City, with its population approaching 300,000, was running out of fresh water. The response of the city fathers was to authorize the construction of a reservoir-aqueduct system that would bring water from Westchester County 41 miles north onto Manhattan Island through an elaborate series of dams, tunnels, overpasses, and pipes carrying 35,000,000 gallons of water daily to a receiving reservoir in what is now part of Central Park. When the project went operational in 1842 it rivaled the Erie Canal as the largest American engineering project to date.[14]

Not coincidentally, the chief engineer of the Croton Aqueduct project was John Jervis, who had come of professional age two decades earlier on the Erie Canal. His chief assistant was Horatio Allen (Columbia College 1823) while Jervis's successor in 1849 as chief engineer was Alfred W. Craven (Columbia College 1829.)[15]

However important these water-related projects in the history of American engineering may have been, two other undertakings, both traversing the immensity of the American landscape, superseded them in size and technical complexity. Railroads—steam-propelled vehicles that moved on tracks—were an English invention of the 1820s. But nowhere in the world did their potential become more immediately apparent than here. In the 1820s the United States encompassed some 1,600,000 square miles, most of which was approachable only by foot. By the 1850s, with additional land wrested from Mexico in 1847, the country was a 3,000 x 1,500 mile hexagon with three north–south mountain ranges obstructing the east–west movement of goods and peoples.[16]

Constructing a continent-spanning railroad system required private capital, governmental assistance, and manual labor, most of the latter provided by an immigrant workforce. Also required were organizational skills and technical knowledge of a high order. While most of that technical knowledge in the early days of railroading was acquired on the job, those who brought some formal training in mathematics and the physical sciences were welcome. One such college railroader was Columbia College graduate, Horatio Allen (1823).[17]

In 1828, Allen, then twenty-one and working on the construction of the Delaware and Hudson Canal, was selected by his employers to purchase a steam locomotive in England and bring it to Pennsylvania. Once back in Honesdale, Allen assembled the locomotive, called the Stourbridge Lion, and was at the controls when it became the first steam locomotive to operate in America. Allen went on to a career in railroading and in the manufacture of marine steam engines. Both he and his son served as trustees of Columbia College.[18]

Local institutional credit for telegraphy, another transformative technological enterprise of the second quarter of the nineteenth century, goes not to Columbia but to the newly opened New York University, where Yale-educated Samuel F. B. Morse was teaching in 1838. There he and a colleague invented a battery-operated instrument capable of transmitting electromagnetic signals through wires. Within fifteen years much of the East Coast would be connected by telegraph lines, bringing almost instantaneous communication across hundreds of miles. Fifteen years later America and Europe were in telegraphic contact.[19]

With this "communications revolution" came the creation of an entirely new form of technically demanding employment, telegraph operator. Young men of technical inclination, a capacity for memorizing Morse Code, and manual dexterity found themselves in demand. Those operators with an inventive bent—the Scottish immigrant Andrew Carnegie and the New Jersey–reared Thomas A. Edison—proposed modifications that brought them to the attention of supervisors and won promotions to more technically demanding positions. Carnegie soon struck out on his own into bridge-building and steel processing, and Edison into electrical power. The industrial agenda of the late nineteenth century had been set.[20]

Another industry, mining, owed its midcentury prominence less to technological breakthroughs or inventions than to geopolitical developments. Iron mining in America began in the seventeenth century, and coal in the eighteenth, though neither was a major employer or nexus of technological innovation. This changed in 1848, following a successful war with Mexico, when the United States took possession of another 800,000 square miles of territory in the southwestern quadrant of North America (the future states of New Mexico, Colorado, Arizona, Nevada, and California) previously claimed by Spain and then Mexico. No sooner had this vast expanse of land been ceded to the United States than gold was discovered in northern California. Thousands of Americans moved west, joined by thousands of immigrants, where together they set about extracting the gold, silver, copper, and lead that lay below.[21]

The railroads, telegraphy, and mining businesses shared several characteristics: the need for financing; organizational complexity; and workforce differentiation. This last concerns us here. All the emergent industries required a labor force with technical skills not readily acquired by traditional apprenticeships or learned on-the-job in canal-building or as telegraph operators. What these new industrial enterprises clearly needed was a portion of their workforce possessed of skills that presupposed formal instruction in science and mathematics at a level comparable to that available to young Americans at mid-nineteenth century entering careers in law and medicine. Less clear was where such instruction could be had.[22]

THE STATE OF THE COLLEGES—THE COLONIAL NINE

A consideration of the scientific and technological competences of early American colleges might begin by pointing out a common misconception about them: they were not primarily theological seminaries. This is not to dispute that all of the prerevolutionary colleges—"the Colonial Nine" or the fifty or so other colleges founded before 1830—were in a broad sense "church schools." They nearly all were founded with denominational purposes (e.g., Congregational, Presbyterian, Anglican, Baptist, Dutch Reformed, Methodist) in mind, typically limited faculty to those of the school's denominational persuasion, and produced substantial numbers of graduates who became ministers within that persuasion. Rather, it is to insist that the instructional fare of these colleges was secular in nature, and, while primarily focused on providing the rudiments of a classical education, not closed to the study of science or mathematics.[23]

That said, attention paid to mathematics and science—the latter then termed "natural philosophy"—varied from college to college, with none of the first four colleges—Harvard (1636), William and Mary (1693), Yale (1701),

Princeton (1746)—devoting more than a modest proportion of instructional resources to either. In 1708, one Harvard tutor (of four) was responsible for all instruction in mathematics and science, which is to say astronomy. Only one of its half-dozen mid-eighteenth-century professors, John Winthrop, was engaged in scientific investigation. Yale, founded in 1701, did not get around to appointing a professor of mathematics and natural philosophy until 1770. Neither William and Mary, founded in 1696, nor The College of New Jersey (later Princeton), founded in 1746, paid more than passing early attention to mathematics or science.[24]

Thus, the possibility of becoming the most scientifically oriented of the early colleges presented itself to the founders of King's College in 1754. That the college was located in New York City, the most religiously diversified of colonial cities, and governed by a board of governors drawn from the ranks of political officialdom and the clergy drawn from the city's four major Protestant denominations, furthered an openness to scientific instruction and inquiry. Also promising was the decision to appoint the Reverend Samuel Johnson as the college's first president. Among mid-eighteenth-century Americans, Johnson was one of the four best-known American intellectual investigators in the wider Atlantic world (Benjamin Franklin, Jonathan Edwards, and John Winthrop the others) and was conversant with the scientific writings of Isaac Newton, Robert Boyle, and John Locke, which he encountered as a tutor at Yale.[25]

Johnson's vision for King's College, as conveyed in a spring 1754 announcement of its opening, reinforced the prospect of a science-friendly environment:

> Lastly, . . . it is further the design of this college to instruct . . . in the arts
> of numbering and measuring, of surveying and navigation, of geography
> and history, of husbandry, commerce and government, and in the knowl-
> edge of all nature in the heavens above us, and in the air, water and earth
> around us, and the various kinds of meteors, stones, mines, and minerals,
> plants and animals, and of everything useful for the comfort, the conve-
> nience and elegance of life, in the chief manufactures relating to any of
> these things.[26]

FIGURE 1.1 Samuel Johnson (1696–1772), first president of King's College (1754–1762). Painting by unknown artist hangs in King's College Room, Columbia University.

Source: Columbia University Archives.

True to this vision, Johnson secured as his first professorial appointment twenty-seven-year-old Daniel Treadwell, a protégé of Harvard's John Winthrop, who joined the faculty as professor of natural philosophy and mathematics. Unfortunately for the future of science at King's College, Treadwell died three years later in 1760. But even then, Johnson secured as his successor Robert Harpur, a twenty-three-year-old Glasgow-trained mathematician. "He is indeed a Presbyterian," Johnson assured one of his concerned Anglican governors-trustees, "but I think from what has yet appeared, he will do very well."[27]

Seven years into his presidency, Johnson tired of his responsibilities even as his trustees tired of him. In 1762 they hired the Oxford-trained twenty-seven-year-old Reverend Myles Cooper as professor of moral philosophy. Johnson took the hint and retired a year later. Among Cooper's first actions as King's College's second president was to downgrade the instruction of mathematics by demoting Harpur from professor to tutor, whose instruction was thereafter available to King's College students only for an additional fee. The "Oxfordizing" of the King's College curriculum proceeded, and Johnson's emphasis on science and mathematics was displaced by what Cooper, who fancied himself a classicist and poet, called "polite learning." What science instruction remained at King's College, as the incoming Alexander Hamilton discovered to his chagrin, came from Dr. Samuel Clossy, of the College's medical faculty. It would be one hundred years and eight presidents before the College would have a president who showed the slightest interest in science.[28]

For all its indifference to science, King's College produced several graduates who went on to make a place for themselves in the annals of American technology. Robert R. Livingston (1765) provided the financial backing for Robert Fulton's steamboats and John Stevens (1768) also became a steamboat developer. Alexander Hamilton (King's College 1773–1775), an artillery officer in the Revolutionary War, authored the *Report on Manufactures* (1791) and promoted hydropower on the Passaic River. All acquired an interest in technology despite their collegiate training. This may explain why the Stevens family decided not to leave their father's estate to his alma mater in New York but instead to the founding of Stevens Institute of Technology in Hoboken.[29]

King's College, having passed on the chance to distinguish itself among its colonial peers for its commitment to the sciences, escaped the added fate of seeing that distinction going to those colleges founded in its immediate wake. Neither the College of Philadelphia (later, the University of Pennsylvania), the College of Rhode Island (later, Brown), Queen's College (later, Rutgers), nor Dartmouth did much with science in what remained of the pre-Revolutionary period.[30]

TECHNOLOGY AND THE COLLEGES
OF THE EARLY REPUBLIC

According to Stanley Guralnick, a close scholar of the subject, American antebellum colleges showed little interest in science before the 1820s. Most colleges had only one instructor charged with scientific instruction, while the two largest, Yale and Harvard, had two. By the 1840s, Yale still had only two professors of science, while Princeton had three and Harvard four. Of the presidents of the leading antebellum colleges, only Jeremiah Day, the mathematician president of Yale (1817–1846), and the long-serving president of Union College (1804–1866), Eliphalet Nott, were men of science. In such company, antebellum and decidedly belle-lettrist Columbia College fit nicely.[31]

William Samuel Johnson (1787–1801), the third president of Columbia College (the new suitably republican name acquired in 1784), a lawyer and recovering Loyalist, was far less interested in science than his father, the first President Johnson. But this was also true of his next six successors, the no-show minister Charles Henry Wharton (1801), Episcopal Bishop Benjamin Moore (1801–1811), minister William Harris (1811–1829), newspaperman William Alexander Duer (1829–1842), classical scholar Nathaniel Fish Moore (1842–1849) and Charles King (1849–1863), "another gentleman in want of a situation."[32]

From the 1780s until the 1850s, the Columbia College faculty had two positions reserved for men of science. The first was a professorship in mathematics (sometimes combined with astronomy), held successively by John Kemp (1784–1812), Robert Adrain (1812–1823), Henry James Anderson (1825–1843), and Charles William Hackley (1845–1857). The most impressive was Robert Adrain, whom historians of science describe as one of the brilliant mathematicians of the early nineteenth century. By all accounts, he was also a lovable teacher, but one that the Columbia trustees let be hired away in 1823 by Rutgers and then the University of Pennsylvania, where he served a rocky two years as vice provost. When he returned to New York in 1828 after a falling out at Penn, he was reduced to providing instruction to young boys at the Columbia Grammar School. Hackley, on the other hand, was recognized from early on as an ineffectual teacher and was terminated in 1857 when it was discovered that his students could not compute simple interest.[33]

The only other science slot at antebellum Columbia was the professorship in natural philosophy (appended by whatever other subject(s) fit the incumbent's fancy). Its successive occupants through 1820 were Samuel Latham Mitchill, professor of natural philosophy, natural history, and agriculture (1792–1800); James Stringham, professor of natural philosophy and chemistry (1802–1812); and John Griscom, professor of natural philosophy and chemistry (1813–1820). The Edinburgh-trained physician, Mitchill, was the best known and has been variously described by his biographer as "perhaps the most versatile man in science in his time." Washington Irving, given the professor's claims to expertise on so many scientific fronts, thought him "spread more than a bit thin." Mitchill left Columbia in 1800 upon his election to Congress. He later figured famously in an 1818 New York City court case, *Maurice v. Judd*, in which he provided expert testimony to the fact that a whale was a mammal and not a fish.[34]

THE VERSATILE "JEMMY" RENWICK

"Taking notes with Renwick," Columbia College sophomore George Templeton Strong recorded in his diary on October 12, 1835. "His room puts one in mind of Virgil's vision of Tartarus, and it smells like everything horrible." And so it smelled for the thirty-three classes of Columbia College students for whom "science" meant "Jemmy" Renwick.[35]

Born in England in 1792, James Renwick came to New York as a young boy, but not too young to enroll in Columbia College at age twelve and to graduate at sixteen. An apprenticeship in his father's import business, followed by a commission as engineering officer in the New York militia during the War of 1812, followed by a stint as the personal secretary to Washington Irving, suggested an impressive if unfocused career start.

In 1816 he was elected to the Columbia College board of trustees. Three years later, now married to a niece of Washington Irving and the father of three young boys, Renwick watched as the Panic of 1819 bankrupted the family business and put his economic survival in doubt. Fortunately, the resignation of John Griscom as professor of natural philosophy the following year led to his being offered the position. "I am heartily glad that James Renwick is snugly nestled in the old college," Irving wrote to the professor-elect's equally relieved brother-in-law. Of his ex-secretary's new position, Irving added: "[it] is a safe harbor of life and a very comfortable and honorable one."[36]

To Renwick's credit, he did not turn his professorship into the comfy sinecure that Irving predicted. Instead, for thirty-three years as professor of chemistry and experimental philosophy he attempted, as the historian Edwin Layton described his career, "to foster the application of science to technology." He did this in his teaching, public lectures, and in several books, most significantly in *The Elements of Mechanics* and *Applications of the Science of Mechanics to Practical Purposes*. Other books came from his lecture notes, including his *Treatise on the Steam Engine* and *First Principles of Chemistry*. His publications extended to translations of French technical treatises and, later, a series of scientific biographies. He became the first Columbia professor to provide services as engineering consultant: to John Jervis's Delaware and Hudson Canal; to Jervis again on the design of the Mohawk and Hudson Railroad; to the Navy on the prevention of boiler explosions; to Rochester millers on rerouting the Genesee River; and to commissioners of the Webster-Ashburton Treaty between Maine and New Brunswick, on topographical measurements.[37]

Renwick taught courses in virtually every field open to scientific instruction, including chemistry, geology, and physics, with occasional courses on steam engineering and ballistics. Among his students were the aforementioned engineers in all but name Horatio Allen (1823), Alfred F. Craven (1829), and William Mitchell Gillespie (1834), the first professor of civil engineering at Union College and an expert on road construction. His three Columbia-trained sons all went on to technical careers: Henry (1833) became a patent lawyer; James Jr. (1836) was one of the century's premier

FIGURE 1.2 James Renwick (1790–1863), professor of experimental philosophy, Columbia College (1820–1853). Introduced engineering course into the College curriculum. Portrait in office of Professor George Deodatis.

Source: Columbia University Archives.

architects, whose work includes New York City's St. Patrick's Cathedral; Edward S. (1839) became an oft-patented inventor.[38]

The bookish George Templeton Strong sometimes characterized him as "Old Stick-in-the-Mud, alias Dr. Blowpipe, alias Prof. Renwick," and compared him unfavorably with Yale's professor of chemistry Benjamin Silliman, a frequent and popular presence on the New York lecture circuit. "Recited with Renwick," Strong reported after a class, "that old ass has absolutely given us some problems to work out in gunnery, just as if we cared a quarter of a continental for parabolas de but en blanc howitzers, breeches, chambers, and so forth." On other occasions, however, Strong acknowledged Renwick's good-heartedness. After classmates set off gunpowder in Renwick's room, Strong wrote in his diary: "Poor Jemmy! They'll tease him out of his senses. He has been as busy as possible ever since to try to find out the contriver of the plot. He won't succeed, but it is too bad to tease him this way, for he behaves very civilly to anyone who behaves so to him." On another occasion when Renwick was being put upon by disruptive students, Strong acknowledged that he bore it "with Christian fortitude."[39]

It is not the mixed reviews he received from students that account for the modest place Renwick has been assigned in the early history of American academic science. He is conspicuously absent from the contemporary panoply of scientific notables, which includes Yale's Benjamin Silliman, Princeton's and the Smithsonian Institution's Joseph Henry, Harvard's Louis Agassiz, Asa Gray and Benjamin Peirce of Penn, and the Coastal Survey's Alexander Dallas Bache, who together comprised what Bache self-mockingly called the "Scientific Lazzaroni, a mutual admiration society" that advanced the collective cause of nineteenth-century American science and scientists. They did so in part through the creation of the American Association for the Advancement of Science in 1847 and wartime lobbying that led President Lincoln to create the National Academy of Sciences in 1863. For such scientists, Renwick's versatility, but even more his interest in applying science to problems of his day, rendered him suspect. In this respect he was seen much like his contemporary, Mathew Fontaine Maury, the director of the Naval Observatory and author of the first textbook in oceanography, *The Physical Geography of the Sea*, as not sufficiently exacting and perhaps too enthusiastic in their utilitarianism. Historians of science, prior to Layton, have been slow to challenge this dismissive contemporary judgment.[40]

Unfortunately for Renwick, this negative assessment of his career was shared by Columbia's most energetic and long-serving trustee, Samuel F. Ruggles, and his son-in-law, George Templeton Strong, who joined the Columbia board in 1853. It was to Ruggles, who summered in Massachusetts, and to Strong, that Harvard's professor of mathematics and astronomy Benjamin Peirce rendered his judgment on the state of science at Columbia. "Peirce holds," Strong recorded in 1852, "that professors may be prevented from degenerating into drones, like Renwick, by requiring of them to accomplish something every year or every six months, making it a condition of holding office that at certain periods they produce some essay, memoir, or investigation in their respective departments." It was on this basis that Ruggles, with Strong's help, engineered the sixty-year-old Renwick's retirement in 1853, to replace him with a younger man who had the full backing of Peirce and the other Lazzaroni. Renwick in retirement remained active, publishing biographies of David Rittenhouse, Robert Fulton, and Count Rumford, plus the *Life of DeWitt Clinton,* until his death in 1863. On the seventy-fifth anniversary of the founding of the School of Mines, in 1929, the university created the James Renwick Chair in Civil Engineering, to honor its otherwise unheralded early exemplar of the American "scientific practitioner."[41]

THE WOLCOTT GIBBS AFFAIR

The young man Ruggles and Strong had in mind to lead the Columbia science department was a former Renwick student, Wolcott Gibbs, Class of 1841, who had gone to study chemistry in Europe, returning to take up the professorship of chemistry at the newly opened Free Academy of New York City (later, City College of New York). The campaign for his appointment commenced in 1850 when Gibbs shared with Strong his ideas about a "polytechnic project for a post-Renwick Columbia College." "Strong prospect of the final elimination of Professor James Renwick from Columbia College," Strong wrote on March 9, 1853. "We must make a grand effort for Wolcott Gibbs if that happens."[42]

Renwick's retirement in November put the campaign into overdrive. Ruggles pressed Gibbs's case at board meetings in early 1854, while Gibbs's outside backers bombarded board members with increasingly unwelcome advice as to their obligation to American science. Thanks to these interventions, as well as Strong's full (if one-sided) record of the board's proceedings, leaks to the press, and Ruggles's public rejection of the board decision not to appoint Gibbs, this became one of the most publicly scrutinized non-appointments in American academic history.[43]

In the end, two sets of trustees did in the Gibbs appointment. The first were board members, all but one either Episcopalians (including three clerics) or Trinitarian Calvinists (Presbyterian or Dutch Reformed) who concluded that Gibbs's Unitarianism disqualified him. Strong privately disparaged them as the "fogy and fossil party." A second set of trustees took offense at the outside lobbying on Gibbs's behalf and at Ruggles and Strong aligning with them. "I trust that you will not consider it an impertinence," Peirce had begun his endorsement of Gibbs, but clearly some trustees did. At the board meeting on February 14, 1854, four ballots failed to produce a majority for any of the three candidates, though Gibbs led each time. A fourth candidate was then added to the list, Professor Richard McCulloch, for the past five years Princeton's professor of natural philosophy and a Presbyterian, with balloting set for the March 6 meeting.[44]

When that meeting resulted in four more ballots and still no majority, Ruggles took the case for Gibbs to the public in a pamphlet provocatively titled *The Duty of Columbia College to the Community*. In an imaginative reconstruction of recent events, Ruggles began by blaming his fellow trustees for Renwick's departure:

> We all know, and the public knows how we turned away, year after year, from all entreaties of our late professor, that his department and the reputation of the College might be relieved from these needless embarrassments, till after thirty years' service, he has left us in disgust and despair.[45]

Ruggles then proceeded to charge the opponents of Gibbs's appointment with religious discrimination, in violation of the college charter. Lest the implied threat of state intervention go unnoted, Ruggles closed with the categorical pronouncement that Columbia was "in no sense an ecclesiastical body. . . . The College is a public, not a private institution." Even Strong thought his father-in-law might have overdone it.[46]

On April 3, 1854, with twenty-one members present, McCulloch received eleven votes, Gibbs nine, and Alexander Dallas Bache one, thereby securing the appointment for McCulloch. The first act of the "post-Renwick Columbia College" was to turn its back on the country's best-trained and most promising chemist, and a Columbia graduate to boot. Not surprising that, in covering the Gibbs affair for its Unitarian readers, Boston-based *Christian Examiner* described Columbia College as "good in classics, weak in sciences, with few graduates of distinction."[47]

Gibbs stayed at City College until 1861, when he was nominated for a new chemistry professorship at Harvard. This time his backers carried the day, with Gibbs beating out the inside candidate, Charles William Eliot, who later took a faculty position at the newly opened MIT before returning to Harvard in 1869 as president. Gibbs remained at Harvard, where he achieved international recognition, until his retirement.[48]

As for McCulloch, Strong found him "a feeble-looking, washed-out kind of man." "It is shameful to hear the man we have entrusted with the teaching of all that vast group of sciences, to whom we have committed the education of our students in knowledge of the material works of god." He dismissed McCulloch's inaugural lecture as "defiling his awful subject by the utterance of trivialities and platitudes so slovenly and contemptible." Columbia's trustees had for the moment repulsed the advance guard of scientific meritocracy in favor of local autonomy and theological conformity. As one of the anti-Gibbs majority put it in a subsequent appointments consideration, "a man without scientific name and standing was preferable to one of the highest reputation ... because the nameless professor would work harder to secure reputation than the professor who had already secured it."[49]

MIDCENTURY STIRRINGS

Even as Columbians did not, educators elsewhere had been implementing variations on Gibbs's "polytechnic project for a post-Renwick Columbia College." Thirty years earlier, in 1824, the wealthy upstate landowner Stephen Van Rensselaer opened the Rensselaer School (later Rensselaer Polytechnic Institute, or RPI) in Troy, New York. Along with the United States Military Academy, opened in 1802 at West Point, New York, and which with the arrival of Sylvanus Thayer as superintendent in 1817 focused its curriculum on military engineering, RPI enjoyed a two-decade monopoly on "instructing persons . . . in the application of science to the common purposes of life." That changed in 1847 when both Yale and Harvard received funding to support applied science. In the Yale instance, it was a $10,000 gift (the first of many) from the Connecticut railroad investor Joseph Sheffield to create two professorships in applied science; at Harvard, it was a $50,000 gift from Abbott Lawrence for "an advanced school of instruction in theoretical and practical science," which included a professorship in engineering. Over the next five years, Dartmouth, Brown and the University of Michigan mounted programs in engineering and applied science.[50]

Meanwhile, the formation of a scientific community proceeded apace. The American Association for the Advancement of Science became the principal lobbying agent. The opening of the Smithsonian Institution in 1848 and the appointment of the physicist Joseph Henry as the first superintendent provided gainful scientific employment in the nation's capital, where the cause of science was thereafter aggressively pressed. It was the federal government and not American colleges that provided the first institutional "home" for American science. The Coastal Survey, the Naval Observatory, and later the Geological Survey and the Department of Agriculture made science a career option, not just the avocational pursuit of the independently wealthy or professed otherwise employed.[51]

The occupational commitment to science often came during postcollegiate study in Germany and France which, by the 1850s, hundreds of American graduates undertook. Some of these young men returned home determined to emulate the scientists with whom they had studied abroad, men whose commitment to science seemed of a disinterested purity akin to a sublime religious experience. For these academic innocents abroad, many of whom might earlier have gone into the ministry, a devotion to science became a substitute for religious belief. Others came back with more utilitarian goals in mind, to put their recently acquired knowledge to work developing America's resources. It was the second sort who came knocking on Columbia's door in 1863.[52]

2

FAST START 1864–1889

Work on the new School of Mines building is in active progress.... Who would have thought that the little egg that was laid in this very room twelve years ago could hatch out so big a bird?

—GEORGE TEMPLETON STRONG, DIARY, JUNE 9, 1874[1]

The School [of Mines] is doing an inappreciable service to the country. Every year its usefulness is increasingly felt.

—PRESIDENT FREDERICK A. P. BARNARD, MARCH 2, 1874[2]

Many of the officers and graduates of the School of Mines take serious exception to the views expressed by Professor Burgess ... that instruction and research in engineering and applied science have no place at Columbia as it transforms itself into the prototype of the American University.

—*THE SCHOOL OF MINES QUARTERLY*, 1884, IN REVIEW OF JOHN W. BURGESS', THE AMERICAN UNIVERSITY: WHEN SHALL IT BE? WHERE SHALL IT BE? WHAT SHALL IT BE?[3]

T WO PEOPLE DESERVE credit for founding the School of Mines of Columbia College: Thomas Edward Egleston, Jr., who proposed the school, and George Templeton Strong, who persuaded his reluctant fellow Columbia trustees to adopt "Egleston's dream." Three people facilitated the fast start and early success: Strong again, as the school's chief trustee sponsor for its first decade; Columbia College President Frederick A. P. Barnard, faithful defender during his twenty-five-year presidency; and Charles Frederick Chandler, professor of analytic chemistry and the school's first and longest serving dean.[4]

EGLESTON'S DREAM, STRONG'S AGENCY

The idea of a "polytechnic school" attached to Columbia dates back to 1850 to Wolcott Gibbs and his trustee backers, Samuel B. Ruggles and George Templeton Strong. The idea for a school specifically of mining and metallurgy came more than a decade later from an outsider, one Thomas E. Egleston, Jr. An independently wealthy New Yorker (his father owned an iron foundry and on his death in 1861 left his three sons an estate of $500,000), Egleston followed up four years at Yale (1852) with an extended tour of Europe that culminated in two years of study at the École des Mines in Paris. He returned to the United States in 1861, at the outset of the Civil War, in search of a position where he could continue his studies in metallurgy and mineralogy.[5]

A professorship at an established college, perhaps at his alma mater in New Haven, might have served nicely, but no such offer materialized.

FIGURE 2.1 Thomas Egleston (1832–1900), founder and professor of metallurgy, School of Mines, Columbia College (1864–1892).

Source: Columbia University Archives.

A second possibility was the Smithsonian Institution in Washington, to which in 1862 the thirty-year-old Egleston offered his mineralogy collection and proposed to offer instruction in metallurgy. When no permanent position materialized, he turned his sights on his native New York City. There he approached family acquaintance Peter Cooper with his idea for a mines and metallurgy school to be affiliated with the recently opened (1859) and tuition-free Cooper Union, only to have Cooper reject his idea as too esoteric for his educational purposes. Egleston then brought his twice-rejected "dream of a School of Mining and Metallurgy" to the attorney George Templeton Strong, a fellow Trinity Church communicant and Columbia College trustee. On May 4, 1863, despite his preoccupation with news from the battlefields around "a one-house village called Chancellorsville," Strong made good on his promise to convey Egleston's plan to his fellow trustees.[6]

Strong is best known to American historians for his diary, which he began in 1837 and contributed to almost daily until his death in 1875. It has provided countless social historians and general readers with a four-decade window into the privileged world of mid-nineteenth-century Knickerbocker New York. It has also served political historians with a day-to-day measure of informed Union sentiment about the progress of the Civil War and of President Lincoln, whom Strong admired and served tirelessly as treasurer of the United States Sanitary Commission. But in so much as Strong began his diary while at Columbia College, was interested in college affairs as an alumnus and the son-in-law of a trustee, and for twenty-two years (1853–1875) served as a trustee, it provides academic historians with the most unvarnished view into the inner workings of any nineteenth-century American college and the birth of the School of Mines of Columbia College, later the School of Engineering and Applied Science of Columbia University.[7]

Much of what Strong wrote about Columbia was critical, often the sentiments of a disappointed lover. As a student he laid into his teachers; as a trustee he regularly disparaged fellow board members, his father-in-law excepted, as penny-pinching and sectarian "old fogeys." Disappointment with his alma mater for all the opportunities for improvement squandered or sabotaged by lazy faculty, feckless presidents, or inept trustees is the diary's prevailing motif. There was one striking exception: Strong's unfaltering

FIGURE 2.2 George Templeton Strong (1820–1875), attorney, diarist, and Columbia College trustee (1853–1875); leading backer of the College's School of Mines.

Source: Columbia University Archives.

thirteen-year love affair with the School of Mines, from first hearing Egleston's idea in 1863 to his efforts just weeks before his death in 1875 to cover Egleston's expenditures—made "without the least authority"—on behalf of the then thriving School.[8]

"I think the board will agree to an experimental beginning on a small scale," Strong wrote on October 24, 1863, regarding the board's response to Egleston's May proposal. Much had happened in the intervening five months. For one, Renwick's successor, the physics professor Richard McCulloch, had decamped after the Battle of Gettysburg to join the Confederate cause. His letter of resignation led to his being struck from the faculty list, and a search for his replacement was initiated. Ogden N. Rood, a thirty-three-year-old physics professor at Union College, was quickly elected by the trustees.[9]

Meanwhile, a late application for the job that went to Rood had come in from Frederick A. P. Barnard, a Yale graduate (1832) who had for twenty-five years taught mathematics and astronomy as well as presided over the Universities of Alabama and Mississippi before effecting a reverse-McCulloch by crossing Union lines in early 1863. Barnard's brother, a general in the Union army, secured him temporary work in Washington. But just as this outspoken Unionist and experienced academic administrator came to the attention of the Columbia board, President Charles King announced his intention to retire, effective June 30, 1864. The board, under public criticism for having in its ranks "Copperheads" unsympathetic to the Union cause, included Barnard's application for a faculty position among its applicants for the presidency. On March 26, 1864, sight unseen, Barnard was elected the tenth president of Columbia College. That the president-elect was a convert to Episcopalianism likely helped his candidacy; that he was a talented mathematician, estimable astronomer, and a charter member of the National Academy of Sciences likely went unnoted.[10]

Meanwhile, plans for a school of mines had gained momentum, especially after board treasurer William Betts secured agreement that Egleston and any additional hires would not be on the college's dime but would be compensated through student fees. This was the arrangement by which Theodore Dwight operated his six-year-old law school, which freed the Columbia board of financial responsibility but also from claims to sharing in the school's considerable revenues. Following favorable board action on October 24, 1863, Strong sought out contributors to a $20,000 fund to get a mining school started. When he approached businessman William E. Dodge, he was told that no contribution would be forthcoming from him or like-minded New Yorkers without evidence of the board's willingness to back the School of the Mines with some of the college's own money.[11]

On December 30, 1863, Egleston was appointed to the nonsalaried professorship of mineralogy and metallurgy in the prospective School of Mines. He in turn recommended Francis L. Vinton, a West Point graduate and student at the École des Mines during Egleston's time there and decorated Civil War officer, who had been breveted out as a brigadier general after sustaining injuries at Fredericksburg. The trustees announced Vinton's appointment as professor of mining on October 5, 1864. Three weeks earlier, following a visit to Schenectady by Egleston, an offer was made to Charles F. Chandler, then a twenty-nine-year-old professor of chemistry at Union, which he accepted in early November. All three were Episcopalians.[12]

On November 15, 1864, in a rented building on the northeast corner of the Columbia College campus then on Madison Avenue and Forty-Ninth Street, with Egleston, Vinton, and Chandler in place, the School of Mines of Columbia College officially opened its doors. Four days later Strong exulted: "The School opened on the fifteenth with twenty students and more are coming!" A week later there were twenty-nine, by Christmas, thirty-three students. Such tangible proof of the school's popularity prompted Strong to lobby his fellow trustees "for an appropriation from the College" and to press his longer-term goal of having the board assume full financial responsibility for the school, including putting the three professors on salary. "If this School can be kept alive through the winter, it will succeed," Strong confided to his diary on November 30, 1864. "And two years hence its supply of educated metallurgists and engineers will begin to add millions to our national resources for the payment of our national debt."[13]

A month later, on January 9, 1865, Strong's high hopes for the School were tempered by Treasurer Betts's urging that, rather than the college taking financial responsibility for the School of Mines, the trustees who supported the School of Mines, along with their outside "associates," might incorporate the school as a separate entity. The proposal "to get the School out of the hands of the College" only weeks after its launching so infuriated Strong that he threatened to resign. He left the meeting determined "to give up all further effort to help the undertaking." Once back home he gave full vent to his disappointment with his fellow trustees:

> I see now why the position of Columbia College is so unlike that of Harvard and Yale and Princeton. It's a consolation to remember that I should have been fearfully overworked for months to come, had not the trustees decided to choke and suppress this movement. But it is a *great pity*, nevertheless, and a public calamity. So much might have been done; so strong a school established, and such proof given that physical science can be taught under church influence.[14]

However inclined some trustees were to "choke and suppress" the School of Mines, its popular acceptance could not be gainsaid. By spring 1865 enrollments climbed to forty-three, nearly matching the number of freshmen in the college, where class sizes had been flat for decades. On March 6, at the urging of a rejuvenated Strong, the board assumed full financial responsibility for the School of Mines, put its professors on salary, and directed student fees to the college treasurer. Part of the decision to put the school on its books, besides trying to convince non-Episcopalian New Yorkers like Dodge to support it, involved buttressing the college's claims to some of the federal funds slated for New York State under the terms of the 1862 Morrill Land Grant Act for the support of agricultural and mechanical colleges, but to no avail. The promoters of the not-yet-opened Cornell University, with greater access to Albany legislators, beat Columbia to the money, and not for the last time.[15]

Even without the Morrill grant, escalating rents on the college's Manhattan real estate holdings ensured that Columbia could afford the three new professorships. In addition, the School of Mines enjoyed the teaching services of already on the college payroll professors Ogden Rood in physics, Charles Joy in chemistry, and William Peck in mathematics, the last assisted by a new adjunct professor of mathematics, John Howard Van Amringe. As confirmation of the board's new commitment to the school, the trustees in May signed a lease on a three-story building just east of the college campus on the southwest corner of Lexington Avenue and Fiftieth Street, hard by the New York Central railroad tracks. It became the first home of the School of Mines of Columbia College.[16]

Enrollments grew apace. By 1867, when the school awarded its first degree in mining engineering (EM) degrees to fourteen graduates, enrollments in its three classes stood at one hundred twenty-three. In a January 1867 diary entry, Strong uncharacteristically allowed: "Better days seen coming . . . thanks to Barnard, Professor Dwight, and the faculty of the School of Mines." By 1874 some two hundred students were distributed among the school's four classes. After a decade in operation, the school's three hundred alumni constituted most of the nation's professionally trained mining engineers—reason enough for Strong to allow himself another

FIGURE 2.3 First School of Mines building, corner of Third Avenue and Fiftieth Street, 1864.

Source: Columbia University Archives.

self-congratulating interrogatory midway through the school's tenth year of operation, when enrollments surpassed those of the college: "Who would have thought that the little egg that was laid in this very room twelve years ago could hatch out so big a bird?"[17]

"THE ADMIRABLE FREDERICK"

In February 1864, plans for the School of Mines well underway, Frederick A. P. Barnard became a candidate to succeed Charles King as president of Columbia College. Strong, who privately acknowledged an interest in the position, thought Barnard possessed important qualifications, not least having "rank in science and letters." But what "absolutely disqualified him,"

Strong wrote, was "his incurable deafness." He allowed, however, "the board might think otherwise." Indeed it did on May 18, 1864, when, after Strong withdrew his own candidacy and in the face of some trustees' objections to appointing any scientific man president of Columbia, the board elected Barnard the college's tenth president. Ten days earlier Strong had confided to his diary that such an outcome "would be a disaster."[18]

Nor did Strong's first personal contact with the new president prompt a reconsideration. "Barnard is manifestly a thoughtful, judicious, earnest, kindly man, of high principle, intense loyalty, and great practice in all educational questions," Strong acknowledged upon meeting him on June 9, 1864. "But he is very deaf, and his deafness will seriously interfere with his official duties, and he has no presence to help him through. . . . He seems physically incapable of governing a disaffected, disorderly corps of students." Four months later, Strong still thought "he is likely to prove a failure."[19]

Yet a month later, two weeks before the School of Mines was to open, Strong registered a complete change of mind. "Barnard's administration as president of the College is thus far most successful, in spite of his deafness. I predict he will prove the most efficient president we have had for thirty-five years." Strong cited as an instance of Barnard's singular tact: "Professor [of Greek] Anthon likes him and speaks of him with respect, though he has never before had anything good to say of an official superior." What clearly won over Strong was Barnard's immediate identification with the School of Mines and his stated determination to join Strong in tending to its care and feeding.[20]

If such early efforts on behalf of the School of Mines endeared Barnard to Strong and other science-friendly trustees, among them Samuel Ruggles, Lewis Rutherford, Cornelius Agnew, and John Torrey, they put the "anti-science" contingent, including several members of the clergy and regularly a bishop or two, on the defensive. The president's willingness to expend college funds on Mines in excess of the school's revenues also worried board members who saw their principal responsibility as growing the college's already considerable endowment. A third contingent consisted of trustees so protective of the classics-oriented School of Arts (aka "The College") that they saw the success of any other part of the institution, particularly a

FIGURE 2.4 Frederick A. P. Barnard (1809–1889), scientist and tenth president of Columbia College (1864–1889); vigorous advocate for the School of Mines.

Source: Columbia University Archives.

part that competed for the limited supply of would-be undergraduates, as threatening what they considered the heart of the institution.[21]

Barnard did little to assuage the concerns of these "College Firsters." It did not escape their notice that the president's two sons enrolled not in the college but in the School of Mines. Augustus (1868) joined the school's second class, his younger brother Frederick (1872) joined the sixth. But the president's public pronouncements put them still more on their guard. Two years into his Columbia presidency, Barnard published a widely circulated statistical analysis of college attendance in the northeast, provocatively titled "On the Declining Popularity of Collegiate Education," in which he noted a two-decade decline in college attendees as a share of its total population, accompanied by a growth in the number of colleges. Barnard concluded that the northeast had too many colleges and that some of the more established colleges should consider abandoning undergraduate instruction altogether in favor of graduate and professional instruction. Among colleges especially well placed to do so, he suggested, were those located in large metropolitan centers engaged in professional instruction. Nobody on the Columbia board doubted what he had in mind. Barnard's reputation as a "College skeptic" became fixed in the minds of those trustees and alumni for whom the under-graduate, classics-oriented, AB-granting College *was* Columbia.[22]

THE PROTEAN PROFESSOR CHANDLER

The third force behind the fast start of the School of Mines was the profes-sor of analytical chemistry and dean, Charles Frederick Chandler. Indeed, were one of an Emersonian bent, casting the early history of Columbia's School of Mines as "the lengthened shadow of a single man," the man would be Chandler. From his 1864 faculty appointment and election as dean the following spring to his forced resignation from that post thirty-three years later in 1897 (on terms which had him stay on as chairman of the Uni-versity's chemistry department) until his retirement from Columbia in 1910, for its critics and champions alike, personified the school.[23]

The first American in this line of Chandlers had migrated to Massa-chusetts in 1637. Five generations on, Charles was the elder of two sons of a prosperous merchant connected with New Bedford's whaling industry. He took an early interest in science, chemistry in particular. In 1853, at sixteen, he matriculated at Harvard's recently opened Lawrence Scientific School. Unhappy with the elementary level of instruction in chemistry there, he sought advice on finding more advanced training in Germany from a visitor to Harvard, Union College Professor Charles A. Joy, who had just returned from Gottingen, PhD in hand. Joy encouraged Chandler to follow suit and recommended him to his Gottingen mentor, Professor Frederick Wohler, then Germany's leading chemist. A fund raised by New Bedford business-men and free passage from New York to Antwerp accompanying a consign-ment of whale oil in the summer of 1854 underwrote the venture.[24]

A gift to Wohler of American minerals eased Chandler's introduction and gained him admission to Wohler's laboratory, where a dozen Amer-icans were at work. After a year at Gottingen he moved on to Berlin to work with another internationally recognized chemist, Heinrich Rose, and through Rose entered into the occasional company of Berlin's leading scientist, Alexander von Humboldt. In the fall of 1855 Chandler returned to Gottingen, submitted his dissertation titled "Miscellaneous Chemical Researches," secured a waiver of his final examination, and set sail from Bremen for America, with a newly minted Gottingen PhD among his baggage. At age twenty, he had become one of the handful of Americans possessed of the credential that would spur the rapid professionalization of American scientific and academic life.[25]

Once stateside in late 1856, Chandler briefly took up employment with his New Bedford sponsors, providing analyses of various blends of whale oil. In January 1857, an offer of an instructorship in chemistry at Union College came from Joy and sent him to Schenectady, New York. Chandler learned on arrival that the salary for his job had been eliminated, but an unfilled position as janitor remained open. Chandler took it and impressed both Joy and the college's aging president Eliphalet Nott with his energy. On Joy's departure in 1857 to Columbia College, Chandler assumed his professorship and full responsibility for instruction in chemistry (and geology). In 1860 he published *Manual of Qualitative Analysis* to be used for instructional purposes.[26]

When the Civil War broke out in the spring of 1861, the still single twenty-five-year-old Chandler did not seek a commission. He remained at Union. In the spring of 1864 Thomas Egleston traveled to Schenect-ady to examine Union's mineral collection and meet the college's young enterprising chemistry professor. The encounter led six months later to an offer from the Columbia College trustees to join Egleston and Fran-cis Vinton as the third member of the founding faculty of the School of Mines of Columbia College. Despite the proviso that his salary, like those of his colleagues, was to be derived "wholly from student fees,"

FIGURE 2.5 Charles F. Chandler (1836–1925), professor of chemistry, School of Mines (1864–1910), Columbia College (1877–1896), Columbia University (1896–1910); the first dean of School of Mines (1865–1897). 1904 portrait property of Columbia University.

Source: Columbia University Archives.

the twenty-seven-year-old Chandler promptly accepted. In so doing, he became part of what had already been a substantial migration of ambitious and once-maritime-connected New Englanders to the newly commercially ascendant New York City.[27]

Four months after classes opened in November 1864, Chandler was elected by his colleagues to be the School of Mines' first dean. Neither Egleston nor Vinton had any interest in the administrative post, whereas Chandler, never one to pass up a thousand-dollar stipend, enthusiastically took up its duties. Two years later he also accepted an appointment as professor of chemistry at the New York College of Pharmacy, located on Washington Square, where for the next forty-four years he offered instruction in chemistry two evenings a week; he also off and on filled in as acting president of the college. Chandler often boasted of having taught most of the city's apothecaries.[28]

In 1872, Chandler added to his multiple academic appointments adjunct professor of chemistry and medical jurisprudence at Columbia-affiliated but still largely autonomous College of Physicians and Surgeons. For the next twenty-five years he taught P & S students five afternoons a week, from 4:00 to 6:00 P.M., in the P & S building on Twenty-Third Street. In 1877, upon Joy's retirement from Columbia, the heretofore two professorships of chemistry (Joy's in chemistry, Chandler's in analytic chemistry) were consolidated into one with Chandler appointed professor of chemistry, a position he held until his retirement in 1910. When Chandler inquired about the possibility of receiving both salaries, President Barnard demurred, though he agreed thereafter that Columbia should pay the salary of Chandler's personal assistant. Until President Low intervened in 1896, Chandler had for two decades been drawing four professorial salaries and taught a minimum of 30 hours a week at four different locations. If not the most thinly spread, he was—at $20,000 a year—unquestionably the highest paid academic in America.[29]

Remunerated public service also beckoned. In 1866 Chandler was appointed consulting chemist for the newly created New York City Board of Health with a salary of $2,000. Useful reports followed thereafter about the dangers of corrupted milk and kerosene being sold in the city. He also assured a concerned public as to the safety of the water from the Croton Reservoir. In 1873 he was appointed to a four-year term as Board of Health president. Although his often high-handed actions—including a predawn unannounced leveling of the fetid Washington Square Market in 1873—put him at odds with storekeepers and city officials, he was reappointed to a second term in 1877. Four years later Tammany opposition and lukewarm support from Republicans on the city council cost him a third term. Thereafter he stayed involved in public health as a member of the New York State Board of Health and chairman of the Charity Board.[30]

Chandler's location in New York, organizational skills, boundless energy, and outgoing personality assured him a leading role in his discipline. Along with his younger brother William, a chemistry professor at and later president of Lehigh, he founded and edited *The American Chemist*. The Chandlers used their journal to organize the first meeting of the American Chemical Society in 1876. In 1881 Charles became the society's sixth president.[31]

Chandler's published work consisted of some thirty-four papers and reflected his shifting scientific interests from analytic chemistry and mineralogy to issues related to public health and epidemiology. As his time became more and more taken up with industrial consulting and providing expert testimony, his scientific output slowed and ceased by the 1880s. Nothing he published had much impact on the fields of analytic chemistry or mineralogy, and his designation as "the father of American industrial chemistry" reflected his efforts to promote American chemistry more than anything he did as a scientist or engineer. His Columbia colleague and friend the botanist Marston Bogert acknowledged that his election to the National Academy of Sciences in 1874 recognized his public presence as a man of science rather than any scientific breakthrough in chemistry. Another otherwise admiring biographer concurred that he "contributed little or nothing to theoretical research."[32]

And what of Chandler as the School of Mines dean? For his first quarter-century, he performed admirably. He handled paperwork with dispatch, aided by a secretary of his employ, and dealt with students and colleagues with skill and tact. He oversaw three moves the school made in its never ending search for additional space on a cramped campus hemmed in on four sides by commercial activity. Part of his early success as an administrator can be attributed to the simple fact that he oversaw what was from its founding and for fifteen years thereafter Columbia's fastest growing and most successful academic unit. The early School of Mines more than paid its early way, which could not be said of the only slightly larger College of the Arts, or later, after its founding in 1880, of the School of Political Science. Also contributing to Chandler's early success was that he got on famously with President Barnard. They took up their respective positions at Columbia within days of each other and for twenty-five years worked in concert. Their being men of science in an academic community where science had yet to establish itself provided a natural bond. So did the fact that each was an outsider with no ancestral claims to membership in the city's Knickerbocker establishment, which founded and governed the College into their day.[33]

Barnard and Chandler also shared a skeptic's view of the educational value provided by the traditional undergraduate college. Each favored change over preservation. Chandler's skepticism with the status quo came naturally from his own educational experience, one unsatisfying year at Harvard's scientific school, which he had sought out as an alternative to Harvard College, and then an extended and rewarding exposure to two of Germany's greatest universities. Barnard's agnosticism stemmed from his own traditional academic experience as a Yale undergraduate and tutor and a quarter century of teaching and administering at the rough and rowdy Universities of Alabama and Mississippi. Along with a minority of the Columbia trustees, Barnard and Chandler were continuously on the lookout for new ways of structuring higher education and making Columbia a leader in that effort.[34]

Barnard and Chandler joined forces in the early 1870s to expand the offerings of the School of Mines more broadly into engineering and the natural sciences, to promote graduate studies, and to appoint several of the school's first PhDs to the Mines faculty. They also fought to secure for Mines students equal standing on campus. When some trustees first questioned the rapid growth of the school, Barnard and Chandler successfully defended it. But just as Barnard's early effectiveness as president would be compromised by years of struggling with trustees on too many fronts, Chandler's visibility and ubiquity laid him open to criticism from trustees and disgruntled alumni and later most decisively from Barnard's successor. But all through the first two decades the School of Mines reaped the benefits of an energetic and respected dean working hand in glove with an effective contingent of science-friendly trustees and an encouraging scientist as president. This triumvirate could not last forever, and once broken up, is yet to be reconstituted. Still, it was great while it lasted.[35]

TRUSTEE PUSHBACK

The resentment generated by the early successes of the School of Mines first surfaced publicly at the trustee meeting on March 3, 1874. Board treasurer Gouverneur Morris Ogden, on behalf of a majority of the Site Committee, moved to adopt a resolution regarding Dean Chandler's request for additional space for the School of Mines. Strong read the Ogden resolution to mean that "the growth of the School be stopped should the School outgrow its proposed enlarged accommodations and that we would do no more for it in any event." A lively debate ensued, with President Barnard the first to speak in opposition to Ogden's resolution. Supporters, according to Strong, then "let off one or two anti-science speeches" and the resolution seemed "sure to pass," at least until he had his say on the matter.

> I spoke for about three minutes about my willingness to limit the number of School of Mines students by raising the standard for admission, and about my unwillingness to be a party to putting in our minutes a resolution deliberately and avowedly dwarfing either the undergraduate corps, the Law School, or the School of Mines. I said I thought such a resolution without precedent in the history of any American college.

The resolution was defeated, Strong privately allowing that "the loss was due to my little speech."[36]

President Barnard, who oversaw the preparation of the board minutes, assigned himself a prominent role in the outcome, taking issue with "the suggestion of the Committee that the creation of the School of Mines has been prejudicial to the prosperity of the College proper." Not so, the president insisted, attributing the drop in college enrollments to the success of Harvard's new elective course in attracting New Yorkers who would have attended Columbia. As for the School of Mines students, Barnard pointed out, more than two-thirds were nonresidents of New York and many were college graduates. "The School, therefore, is not a competitor with the College." Unable to leave it there, the president laid into his fellow committee members:

> A proposition to arrest the School in its upward progress, a proposition to put deliberately an obstruction in the way of its increasing usefulness, and such a proposition proceeding from its creators and its natural guardians is something so extraordinary that the undersigned is quite unable to understand it.

Understand it or not, board members prepared to champion the School of Mines against any attempts to slow its progress were now on notice that an anti-Mines alliance comprising the board's "anti-science" faction, its grow-the-endowment money managers, and those fearful of the college being "overshadowed" was taking shape.[37]

Checked for the moment by Barnard and Strong, trustees opposed to an expanded School of Mines did not abandon the field. Time favored them. Strong, in the terminal stages of an illness, attended his last board meeting on May 3, 1875. "A dozy meeting of the College trustees began at two o'clock—two hours of Boeotia," he wrote that evening. Twenty-three years a trustee, he was still "finding it amazing how some of us dread and shrink from any contact of the College with the community." His particular irritation was the board's disapproving the use of a room in the School of Mines for occasional meetings of the New York Board of Health, but it prompted his final pronouncement on alma mater: "One would think the newspaper reports of these meetings an advertisement of the School and that we should seek to make the College buildings a center for all reputable scientific and literary organizations of the city." He died two weeks later, depriving his beloved School of Mines of its most effective defender and Barnard of his most dependable ally.[38]

A year later trustee action mandated that the School of Mines curriculum be altered from the original three years to four, making mandatory what had been since the late 1860s an optional preparatory year to the original three-year program. Critics thought its shorter curriculum had given it an unfair competitive advantage over the four-year college. Supporters,

FIGURE 2.6 Members of School of Mines, Class of 1878, outside a mine site in Pennsylvania in 1877.

Source: Columbia University Archives.

Barnard prominent among them, pointed to the fact that entering Mines students were on average at least two years older than those entering the college (nineteen as opposed to seventeen) and were academically more motivated and eager to get careers underway. The addition of a fourth year did have the effect intended by its sponsors of briefly slowing the growth of the school. In 1880, for the first time in sixteen years, registrations failed to go up. A year later the upward trending enrollments resumed when the school introduced instruction in architecture.[39]

Trustee opposition to the School of Mines in the post-Strong era took other forms. At its March 1878 meeting, the president was called upon to address trustee concerns about Mines students whose tuition was waived in exchange for instructional services rendered as faculty assistants, waivers that might be thought of as Columbia's first graduate fellowships. These "free students," Barnard pointed out, represented about 10 percent of each class, the same percentage of nonpaying students in the College, where they had no instructional responsibilities. Those at Mines included several graduates who were earning PhDs and would go on to teach at the collegiate level, three at Columbia. Lest there be any doubt that the president believed Mines students worth the investment, he favorably compared the school's "successes" with the markedly fewer he found among the "free" graduates of the College, using numbers to back up his argument. Barnard was not someone with whom to duel when the choice of weapon was statistics. But he was also deaf to the impact his put-downs had on trustees of the nonquantifying persuasion.[40]

Trustee concerns about paid assistants also turned on their suspicion that professors were slighting teaching duties in favor of outside employment. Egleston, Vinton, and Newberry all regularly advised mining companies during the summers and in term time. Some of these companies in turn began to identify their consultants—and their institutional affiliation—in prospectuses directed at investors. When trustee treasurer Ogden learned of this practice, he secured an addition to the college statutes at the board meeting on May 6, 1878, expressly forbidding faculty from having their institutional affiliation used in any business transaction. Three years later, when Newberry's name and that of "Columbia College" appeared in court filings and newspaper accounts involving a stockholder suit against Colorado Prime Mine, Ogden pounced. Only Barnard's vigorous defense of the moonlighting geologist on the grounds that the company had used his institutional affiliation without Newberry's permission, added to Newberry's fulsome apology, kept the board from firing him. But the message was conveyed.[41]

What influence Barnard retained over the post-Strong board ended in 1879 when he undertook a campaign to bring coeducation to Columbia. Three successive annual *President's Reports* were devoted to the subject, each more forcefully argued than the previous one and read by a majority of the trustees with growing irritation. Upon the publication of the last of his reports, on June 6, 1882, the board voted to eliminate the president's budget for printing. Had support existed among other trustees for admitting women to the college, the School of Mines, the Law School, and the School of Political Science, no one voiced it in the face of opposition expressed by the board's senior members. When faculty—in the person of Professor John W. Burgess, dean of the two-year-old School of Political Science—and students—in the person of student leader and Burgess acolyte, Nicholas Murray Butler (1882)—also rejected Barnard's call for coeducation, they confirmed how out of step (and ahead of the times) the seventy-three-year-old reform-minded president was with prevailing campus sentiments. Barnard's presidency still had another six years to run when he abandoned his public advocacy for coeducation. But its effectiveness was at an end and with it his ability to shield his beloved School of Mines from its adversaries.[42]

Instances in the later years of the Barnard presidency included the Board of Trustees rejecting proposals relating to the School of Mines for no apparent reason other than they had presidential sponsorship. In 1881 Barnard proposed to introduce an electrical engineering course into the school's curriculum. The suggestion came from none other than Thomas A. Edison, then completing his Pearl Street power station and laying plans for the electrification of lower Manhattan. When Barnard took Edison's proposal to his trustees, along with his own backing and that of Dean Chandler, plus assurances from Edison that he would make equipment available for instructional purposes, the board declined to act. Although Columbia's rental income in the early 1880s allowed the college to enjoy a revenue-over-expenditures edge upwards of $100,000 a year, the anticipated start-up cost of $10,000 for such a program was cited as reason for not proceeding. A year later, both MIT and Cornell introduced instruction in electrical engineering, thus sharing the distinction of being the field's founding institutions. The Columbia board waited another six years before approving a course in electrical engineering, in January 1889, at the second meeting after accepting Barnard's resignation.[43]

Another instance of board obstructionism was the refusal to give serious consideration to a Barnard proposal in 1881 that the name be changed to "the School of Applied Science." The president reasoned that "School of Mines" no longing conveyed the school's expanded range of instruction and degrees offered. In the first decade of operations, the school had awarded most of its degrees in mining engineering (EM). By the mid-1870s, however, the degree of PhB began to be awarded to graduates majoring in applied chemistry or metallurgy. In 1877, following Vinton's resignation and the appointment of physicist and civil engineer John Trowbridge, the School of Mines acquired a faculty member whose interests were not primarily identified with the needs of the mining industry but with engineering more generally. The first degree in civil engineering (CE) was awarded by the School of Mines in 1880.[44]

Clearly, the school had become about more than mines and mining. But equally clearly Barnard's proposal to trustees that a name change reflect the new curricular reality fell on deaf ears. Unspecified alumni opposition to a name change was cited for the board's summary rejection of "The School of Applied Science." Egleston was also against a change in the original name. "The School of Mines" it would remain until an alternative was proposed by a president in better trustee favor than Barnard.[45]

A proposal originating with Professor of Geology and Paleontology John Henry Newberry and his assistant, Nathaniel Lord Britton (PhD 1877), that the School of Mines curriculum be expanded to include an experimental course in botany secured trustee approval in 1880. Two years later, when Barnard and Chandler approached the board for a broader mandate for the School of Mines to commence instruction in what Barnard alternately called "courses of an organic nature" or "natural history," a majority of the board, led by Treasurer Ogden, refused. Barnard did not minimize the consequences of this rebuff:

> Until our educational scheme shall have been made as complete in regards to this class of subjects as it is in respect to others, our College cannot be entitled to hold the position which as the undersigned believe, it is easily practicable to secure for it, that of the leading University on the western continent.[46]

FIRST FRUITS: MINES STUDENTS AND GRADUATES

Even as the successes of the School of Mines in the first quarter century engendered opposition to its further enhancement, faculty went about the business of providing instruction for the twelve hundred young men who entered the school between its opening in 1864 and 1888. Enrollments grew steadily for the first eighteen years, then leveled off with annual enrollments ranging between 350 and 375 in the late 1880s.[47]

Admission followed an oral examination administered by the faculty. Beginning in 1866, applicants under nineteen and those lacking adequate preparation were enrolled in a one-year program at the loosely affiliated School of Mines Preparatory School, located at 32 East Forty-Fifth Street, with J. Woodbridge Davis (CE 1878, PhD 1880), headmaster. Completion of the preparatory program ensured admission to the School of Mines. When the curriculum was extended to four years in 1877, the prep year was added at the front end and the connection with the preparatory school ended.[48]

Only about half those who enrolled in the school stayed to graduate. Some of the attrition was a result of academic difficulties, especially after 1875 when the trustees extended the curriculum to four years and lowered the entering age to eighteen. But likely as important was the presence of many young men in a hurry. In a workforce where any formal technical training was a rarity, even a few courses under their belts made School of Mines "dropouts" readily employable.[49]

Contributing to attrition was also the fact that a substantial proportion of Mines students entered with prior collegiate training. Of the first sixty graduates, twenty came with ABs in hand. Many others had prior work experience not found among the younger College students. Not surprisingly, faculty who taught both College and Mines students, among them Ogden Rood, John Henry Newberry, and John Howard Van Amringe, noted the greater maturity and seriousness of purpose of Mines students as compared with those—Rood called them "dunderheads"—of the College.[50]

FIGURE 2.7 School of Mines students with surveying instruments during summer field trip in Connecticut, near later site of Camp Columbia, 1883.

Source: Columbia University Archives.

Early Mines students showed little deference to the students of the older School of the Arts. After eight years of required attendance at commencements where all seven speaking parts went to graduating seniors from the College, Mines students petitioned the trustees in the spring of 1877 for full participation at commencement. They got it. That same year, unhappy with the skimpy coverage in the occasional student publications emanating from the College, Mine students launched *The Miner,* which began publication in the same year as the *Columbia Spectator.*[51]

Even the matter of the high number of academic failures among School of Mines students, proportionally higher than in the College, became a source of pride. "The high average of failures," a recent graduate wrote in the *School of Mines Quarterly*, "furnishes additional testimony, if that were needed, of the fact that the School of Mines does not abound in delicate 'roasts' and that to graduate from the School means something."[52]

What perhaps most distinguished early Mines students from their contemporaries in the College was the eagerness of many of them to stay on after graduation to pursue advanced studies. As early as 1868 a graduate of the School of Mines petitioned the faculty to continue his studies with Dean Chandler. In 1871, a contingent of Mines graduates formulated a proposal that President Barnard presented to the board of trustees calling for the school to inaugurate a program of graduate studies leading to awarding the degree of doctor of philosophy. At that time only Yale among American academic institutions awarded PhDs, having done so on only a dozen occasions since introducing the degree in 1861.[53]

On June 2, 1873, the trustees accepted a doctoral degree plan presented by the Mines faculty consisting of two years of study beyond the baccalaureate, an examination, and the presentation of a thesis. Two years later, in June 1875, the School of Mines awarded its first PhB, in analytic chemistry, to Elwyn Waller (PhB 1870), a student of and teaching assistant to Dean Chandler. To Waller goes the distinction of being Columbia's first PhD, and to Columbia the distinction of being, after Yale, Cornell (1873), and Harvard (1874), the fourth American university to award the degree. By 1882, when Columbia's then two-year-old School of Political Science awarded its first PhDs, the School of Mines had already awarded thirty-six. The permanent beginnings of Columbia as a graduate institution thus properly begin, not with John W. Burgess and the Faculty of Political Science, as some accounts have it, but with the School of Mines and its striving students.[54]

Included among the early PhDs graduating from the School of Mines was the first woman to graduate from Columbia. The circumstances by which Winifred Edgerton came by this distinction in 1886, only four years after the trustees rejected president Barnard's proposal to open Columbia College to women, were unusual, indeed "absolutely exceptional." Upon graduating from Wellesley in 1883, where she excelled in mathematics and astronomy, and a brief stint at the Harvard Observatory, Edgerton returned home to New York hoping to embark on a course of graduate studies in astronomy, making use of Columbia College's new observatory. When approached, President Barnard predictably supported her application, and the recently appointed professor of astronomy John Krom Rees (EM 1878, PhD 1880)

FIGURE 2.8 Winifred Edgerton (Merrill) (1862–1951), Wellesley College (AB 1883) and the first woman to earn a PhD at Columbia, while a graduate student in astronomy in School of Mines.

Photograph the property of Wellesley College Archives.

agreed to take her on as a special student. But the key enabler in securing her admission was the rector of Trinity Church, the Reverend Morgan Dix.

The Edgertons were communicants of Trinity Church and thus Winifred was known to Dix, who by virtue of his ecclesiastical position had a seat on the Columbia board. It was he who secured her admittance by setting the "absolutely exceptional conditions" explicitly intended to avoid a precedent, among them that Edgerton was to work alone in the observatory and not attend classes (a stipulation she seems to have ignored). Her study and research proceeded smoothly and in the spring of 1886 the School of Mines faculty unanimously recommended she be awarded a PhD, to which the trustees agreed without dissent. Degree in hand, Edgerton considered accepting an appointment at Smith College, only to decide against it in favor of marrying James Merrill, an 1890 Columbia PhD in geology. Family responsibilities and her husband's wishes put an end to all thought of a career in science, but not to allowing the School of Mines to claim bragging rights as Columbia's pioneer in the education of women.[55]

A final indicator of the pioneering initiative of these early Mines graduates was how quickly they set about creating a means of communicating with each other and with the larger engineering community. In November 1879 the first issue of *The School of Mines Quarterly* appeared. As its first four editors, all recent graduates, rightly said of this venture into scholarly journalism, they had "no models from which to draw." This and subsequent editions included regular contributions from Mines faculty and graduates. Publication costs were subsidized by the School of Mines Alumni Association, founded in 1882, which appropriated the back end of the *Quarterly* to transact alumni business. Antedating the *Political Science Quarterly* by eight years, it remained in regular publication for three decades before being folded into the *Columbia University Quarterly* in 1909—one more first.[56]

A few Mines students came from the city's same Knickerbocker families that traditionally supplied their sons to Columbia College, but most did not, and fewer were native New Yorkers. Young men from outside the northeast and foreign-born students were almost from the school's opening proportionally more common among Mines students than in the College. Some had recently emigrated from Europe, others came from Central America and the Caribbean. The first Latin American student to graduate was from Cuba, Louis de Souza Barriosa, in 1877. He was followed by the first two identifiably Asian graduates, Yothimosuka Hasegawasa and Nawokichi Matsui, in 1878. Matsui stayed on for another two years to earn his PhD in 1880 before returning to Japan and a professorship at Kyoto University.[57]

Most Mines students pursued a fixed three- and later four-year course leading to a degree in mining engineering. This was done in classes taught by Professors Egleston, Newberry, Vinton, and, with Vinton's departure in 1877, Henry Smith Munroe. Others concentrated their studies under Chandler in analytical/applied chemistry, graduating with a PhB; a smaller segment concentrated in geology (PhB, with Newberry), and a still smaller number in metallurgy (MetE, with Egleston). Beginning in the early 1870s, before Trowbridge's arrival in 1877, courses were offered in civil engineering, with the first CE awarded in 1874. In 1881, with the appointment of William R. Ware, the school began offering courses and awarding a PhB in architecture. By the mid-1880s the school was graduating a dozen architects a year. Unlike most American engineering schools of the late nineteenth century, whose curriculum retained much of the character of machine shops and emphasized "hands-on" instruction reminiscent of the apprentice system by which practical knowledge passed from master to apprentice, the School of Mines favored formal classroom lectures delivered by professors. In so doing Mines reflected its academic setting even as the mode of instruction and the status it gave professors encouraged a larger number of its graduates to take up academic careers.[58]

The mining industry absorbed the largest percentage (42 percent) of the first five graduating classes, with manufacturing and "business" the next (21 percent). Academia ran a close third, accounting for ten of the first eighty-one known occupational outcomes. Later classes were more occupationally dispersed with proportionally fewer academics, though mining remained the largest employer of graduates through the 1880s. In the early 1890s, School of Mines graduates made up over 40 percent of all that industry's college-trained personnel.[59]

The fact that most of the first Mines graduates took up careers in mining resulted in a far greater residential dispersion than that of contemporary graduates of Columbia College, and likely of other eastern colleges. Whereas three-quarters of all Columbia College graduates in the late nineteenth century worked and lived in New York City, fewer than half of School of Mines graduates did, with 40 percent taking up residence outside the Northeast. Colorado, Arizona, and New Mexico claimed nearly as many early Mines graduates as did Connecticut and New Jersey. The mining towns of Leadville, Colorado; Deadwood, Arizona; and Tombstone, South Dakota were home to multiple Mines graduates in the 1880s.[60]

Mines graduates also moved about more than the more professionally fixed graduates of the College or the law and medical schools. A common pattern for a successful mining engineer was to pass his early years in the mining regions of the Southwest before moving to Denver or San Francisco and then, in some cases, returning east to corporate headquarters. A sizable

FIGURE 2.9 William Barclay Parsons (1859–1932) (Columbia College, 1879; School of Mines, 1882), leading engineer and founder of Parsons Brinckerhoff; Columbia University trustee (1897–1932). Pach Brothers photograph portrait.

Source: Columbia University Archives.

number spent extended periods abroad, some in Latin America, others in Asia, and at least one (Thomas H. Leggett, EM 1879) in South Africa.[61]

By and large, Mines graduates went on to professionally successful, financially rewarding or socially engaged careers. Several who did all three also maintained close ties to Columbia as alumni, including five who became university trustees. Frederick A. Schermerhorn (EM 1868, trustee 1881–1906) and Lenox Smith (EM 1868, trustee 1883–1913) stayed in New York and entered into their family businesses. Benjamin B. Lawrence (EM 1878, trustee 1909–1920) went west as a mining engineer before returning to New York as a principal in a mining firm. William Barclay Parsons (1879, EM 1882, trustee 1900–1931) formed his own engineering consulting firm in New York City four years after graduating. The firm later became (and remains) Parsons Brinckerhoff, which during his time undertook some of the world's largest engineering projects, including the original IRT line of the New York City subway system, the Cape Cod Canal, and a substantial portion of the Chinese railway system. Walter Aldridge (1887,

FIGURE 2.10 Herman Hollerith (1860–1929) School of Mines (EM 1879,
PhD 1890), inventor of mechanical tabulator and founder
of Tabulating Machine Company, later merged into IBM.
Photograph the property of the Library of Congress.

Source: Columbia University Archives.

trustee 1922–1928), like Lawrence, began his career in the Southwest before
returning to New York as president of Texas Gulf Sulphur Corporation.[62]

Other early graduates with notable careers in business included Herman
Hollerith (EM 1879, PhD 1889), whose invention of a mechanical tabulating device in the 1880s, which he used in the 1890 federal census, eventually
led him to form his own company. His Tabulating Machine Company in
1924 was one of the companies that merged and became the International
Business Machine Company. Herbert H. Porter (EM 1886) started in mining in Mexico, then returned to New York to found the engineering company of Sanderson and Porter. Francis B. Crocker (EM 1882), in addition
to being a member of the engineering school faculty (1889–1907), became
the founder and cofounder of several companies manufacturing electrical
motors; the last of these, Crocker-Wheeler, became one of the nation's largest manufacturers of electrical products.[63]

At least twenty Mines graduates went on to careers in academia or science. Eleven did so at Columbia (see Appendix for Faculty Roster); another nine filled professorships at Cornell, the University of Wisconsin, Stanford, Hamilton College, Seton Hall University, and the Colorado School of Mines, where two Mines graduates taught. Others assumed professorships abroad at McGill University and Kyoto University.[64]

President Barnard may have been biased in favor of his sons' school in designating proportionally more "successes" among Mines graduates than among those of the College, but nothing in the documentary record suggests that he was wrong.

BURGESS AS NEMESIS

While individual trustees had complaints with their president, some because of his unswerving support of an expanded School of Mines, none openly questioned the very existence of the school. Nor did the alumni, although some, like Frederick Augustus Schermerhorn (EM 1868) and his "Committee of Ten" in 1884, evinced a form of "tough love" in criticizing some of the school's policies. Its 127 recommendations included extending the school year, stiffening admission requirements, and dropping French and German as requirements for graduation. For out-and-out opposition to the school one must look to the Columbia faculty.[65]

A native of Tennessee, Union veteran of the Civil War, Amherst graduate, student at German and French universities, and Amherst professor, John W. Burgess came to Columbia in 1876 with a mandate to establish a graduate school in political science (by which was meant the social sciences more generally) along the lines of the one rejected by the Amherst trustees a year earlier. Once in New York, and with the active support of Trustee Samuel B. Ruggles and President Barnard, Burgess recruited several like-minded Amherst colleagues to form the core of what became in 1880 the School of Political Science, with him as its first dean.[66]

In many ways Burgess and President Barnard might have made natural allies; both were outsiders to New York and Columbia; both were immune to the romantic/nostalgic allure of collegiate life. Each envisioned Columbia as a place for serious adults and no place for faculty unacquainted with the higher scholarship or the world beyond the campus. What put them at cross purposes was the narrowness of Burgess's view of who should partake of his imagined Columbia. Burgess's Columbia systematically excluded Jews, African Americans, and women, while Barnard's explicitly welcomed all three. They broke permanently in 1882 when Burgess successfully led the on-campus opposition to Barnard's three-year campaign to secure the admission of women to Columbia College. Fifty years later Burgess boasted he did so by raising the double specter among alumni of a coeducational Columbia "overrun by women, especially Hebrew women."[67]

Perhaps emboldened by his success besting Barnard on coeducation, Burgess decided in 1884 to publish his own views on the proper ordering of educational institutions. If some thought such pronouncements fell within the president's exclusive purview, Burgess did not. His twenty-two-page pamphlet *The American University—When Shall It Be? Where Shall It Be? What Shall It Be?* (1884) directly addressed the Columbia Board of Trustees, which he later shamelessly characterized as "the most ideal body known to civilization." Much of what he wrote was intended to please them, not least his prediction that the first American university-to-be would almost certainly be under private auspices and located in New York City.[68]

Burgess laid out a university structure very much like the one he encountered as a student at the University of Gottingen. His American university was to be capped by a Faculty of Philosophy, wherein instruction would be offered in the humanities and social sciences, with Faculties of Jurisprudence and Medicine arrayed below, but not, departing from the German model, a Faculty of Theology. He made only the most modest provision for the physical sciences, and hardly any for the natural sciences, placing both under the protective custody of the Faculty of Philosophy.[69]

It was what Burgess explicitly excluded from his American university that matters here:

Although the university should instruct as well as discover and conserve, yet its means and its energies should not be expended upon the mere *pratique* of its subjects.... And not in the natural sciences is its work their application to the exploitation of the wealth of the universe—that is the subject of the polytechnicum; and all of these things are the industrial side of knowledge and instruction, which should take care of itself, or else be taken care of by those immediately interested therein, and has no claim upon the community at large for sacrifice and support.[70]

There was nothing original in Burgess's fulminations about the "the industrial side of knowledge and instruction"; it passed as the conventional wisdom among German academics of his student days. The English scientist Thomas Huxley's 1875 essay "Science and Culture" had even harsher things to say about "applied science," what Burgess meant by "the industrial side of knowledge and instruction."

I often wish that this phrase, "applied science," had never been invented. For it suggests that there is a sort of scientific knowledge of direct practical use, which can be studied apart from another sort of scientific knowledge, which is of no practical utility, and which is termed "Pure science."[71]

Such sentiments had become in Burgess's day commonplace with nineteenth-century America's promoters of "pure" science, among them the physicist Joseph Henry in 1869: "We leave to others with lower aims and different objects to apply our discoveries to what are called useful purposes." Or with Henry Rowland, the first professor of physics at the newly opened Johns Hopkins University, in "A Plea for Pure Science," addressing the American Association for the Advancement of Science in 1883, a year before Burgess's pamphlet: "Commerce, the applications of science, the accumulation of wealth, are necessities which are a curse to those with high ideals, but a blessing to that portion of the world which has neither the ability nor the taste for higher pursuits."[72]

However derivative, the local implications of Burgess's *American University* were immediately apparent to an anonymous reviewer in *The School of Mines*

Quarterly. "He believes," the reviewer wrote of Burgess, "that instruction and research in engineering and applied science have no place at Columbia as it transforms itself into the prototype of the American University." Nor did the reviewer think himself alone in so inferring: "Many of the officers and graduates of the School of Mines take serious exception to the views expressed by Professor Burgess." One letter to the editor from a Mines graduate called the pamphlet "an injurious reflection upon the School of Mines" and another "a covert sneer."[73]

When the editor of the *School of Mines Quarterly* asked Burgess to respond to the critical review and accompanying letters, if only "to correct this misapprehension," he did so meekly: "There is no reference in my pamphlet to the School of Mines, and none intended. My pamphlet is an abstract discussion of the subject of the American university, and contains no reflections upon any specific institution." However placatory in tone, Burgess neither retracted what he wrote or repudiated the critical inferences drawn. Traditionally humanities and newly social science centered, but science leery and technology averse Columbia had its first well-placed apologist. He would not be the last.[74]

3

A CORNER IN THE UNIVERSITY
1889–1929

Thus at one stroke Columbia ceased to be divided into fragments, and took upon herself the aspect of a university.

Our grandfathers attempted it as an advanced or graduate school, but they found themselves at least a half century in advance of general thought and practice. They had to drop their standard of admission and change their methods of work in order to deal with the facts as they then were.

T HE CLOSING DECADE of the nineteenth century and the two that followed witnessed the full transformation of Columbia from the small, provincial, classics-focused college "under church influence" of George Templeton Strong's era to the thoroughly secular and comprehensive national university recognizable today. In 1900 Columbia was the largest and richest university in America, and ten years later, according to Edwin E. Slosson's *Great American Universities,* "Columbia, situated in the largest city, has the best chance to become the greatest of American universities—and it is improving the chance." Moreover, in its young president, Nicholas Murray Butler, it had, with the retirement in 1909 of Charles W. Eliot from the presidency of Harvard, the nation's most recognizable academic leader. For all the promise of a prosperous future for Columbia, while some of its schools were to prosper, others found their place in the university increasingly circumscribed and even contingent. So it was with engineering.[3]

ENGINEERING IN SETH LOW'S "NEW EPOCH"

President Barnard's resignation in November 1888 followed on several years of wrangling with his trustees. Intellectually thriving, competitively aggressive, and ever wealthier, Columbia was an organizational mess. It now included, in order of founding, seven units:

> —the original undergraduate college for men ("College of Arts" or simply "the College");
>
> —a loosely affiliated medical school ("The College of Physicians and Surgeons");
>
> —a law school—still largely autonomous—and physically separated ("Columbia Law School");
>
> —an engineering school, with both undergraduate and graduate programs ("The School of Mines");
>
> —a fledgling graduate program in the social sciences ("The School of Political Science");
>
> —an informal graduate program in the humanities taught by College faculty;
>
> —a half hearted "Collegiate Course for Women" about to be displaced by an affiliated women's college ("Barnard College").

Bringing order to this inherited "bundle of unrelated schools" fell to Barnard's successor.[4]

Seth Low, Columbia's eleventh president, was in many ways a great Columbia president, perhaps the greatest. From today's perspective, he was right on the major social issues: the education of women, which he favored, though for undergraduates in an affiliated Barnard College; openness to Jewish students and to the election of Jews to the Columbia board; making Columbia nonsectarian in fact; commitment to civic improvement and strengthening the university's links to the city. He was also an able administrator, as even his successor Nicholas Murray Butler later backhandedly acknowledged: "He would have been good running a steamship company or a manufacturing firm."[5]

The Lows, who first settled in Massachusetts in the seventeenth century, came to Brooklyn in the 1820s, part of the exodus from New England maritime towns to New York in the wake of the War of 1812. A. A. Low and Brothers became, in the three decades before the Civil War, one of the nation's biggest shipping firms, with investments in the China trade and in clipper ships plying the New York–San Francisco route during the California Gold Rush. The partnership later underwrote railroad development. Already Episcopalians when they left New England, Abiel Abbott Low and his wife sent their only son, Seth, to Columbia College, thus becoming Knickerbockers by adoption.[6]

Seth Low attended Columbia in the late 1860s, just as the School of Mines was getting underway. In the School of Arts, he took no classes with Mines faculty and had little interaction with the older Mines students. Graduation in 1870 was followed by an obligatory stint in the family business, marriage, and increasingly active involvement in civic affairs. In 1881 he was elected to the Columbia board of trustees. His political activities, including two terms as mayor of Brooklyn (1882–1884) and possibly his relative youth at election (thirty-one), kept him from taking a prominent role in board affairs. Although he later credited Barnard with starting much of what he later implemented, as a trustee he did not number among the embattled president's allies. In 1888 Low sold the family business, leaving him at thirty-eight both independently wealthy and open to new responsibilities. On October 7, 1889, he was elected Columbia's eleventh president.[7]

New to academic administration, Low looked to more experienced hands for advice. Unfortunately for the School of Mines, he found the technology-averse John W. Burgess, along with a protégée of Burgess, the up-and-coming professor of philosophy Nicholas Murray Butler (CC 1882, PhD 1884). Low's presidential efforts to organize Columbia in "the aspect of a university" all followed lines proposed by Burgess and Butler. Together, they constituted a formidable force for reordering Barnard's "bundle" of schools and affiliates.[8]

With the School and Faculty of Political Science providing graduate instruction in the social sciences since 1882, the first move by the trustees, at the direction of their new president and his faculty advisers, was to create a parallel School and Faculty of Philosophy, which would provide graduate instruction in the humanities. The thirty-one-year-old Butler was elected its first dean. Each unit was designated a "University" or "Graduate" Faculty, to distinguish them from the undergraduate College of Arts, hereinafter "Columbia College," which would draw its instructors from the two graduate faculties.[9]

Reintegration of the medical school in 1891 and the simultaneous termination of the proprietary arrangements at the law school resulted in both instructional staffs being reorganized as free-standing professional school faculties, each with its own dean. Both enrolled students seeking professional training who were either college graduates or had attended another college long enough not to be interested in attending Columbia College. Indeed, the idea was that both schools would draw a good share of their students from graduates or upperclass transfers from the College, which is what happened.[10]

The matter of organizing the sciences proved to be more complicated and more contested. The position of the School of Mines was enunciated earlier by Professor of Mathematics John Howard Van Amringe, a member of the School of Mines faculty since its founding, but who also taught in the College (and would, in 1896, become its first dean), who wrote in the *School of Mines Quarterly*:

From the beginning of the Graduate Department of the College in 1880 till now, all the instruction in pure science therein given has been given by professors of the School [of Mines]. In the fuller University development that is coming on apace, there is no reason why this School, not the least of the educational glories of Columbia College, should not be acknowledged, as it is the real representative of pure as well as applied science. There is every reason why it should have the direction and control, under the trustees, of the advanced studies for university degrees.[11]

Van Amringe wanted the sciences permanently housed within the School of Mines, which would change its name to something more all-encompassing, either, as Mines faculty proposed in January 1890, "The University Faculty of Science, Pure and Applied," or four months later, "The Faculty of Science," which would "encompass Mathematics and the Natural Sciences."[12]

The inclusion of the natural sciences could be defended on the same historical grounds as the physical sciences, namely that instruction in the natural sciences at Columbia began in the School of Mines. When Barnard and Chandler recommended back in 1883 that Columbia expand its offerings in what the president called "organic nature," they explicitly designated the School of Mines responsible for such instruction. But Barnard's recommendation was tabled by the trustees and not taken up again until after Barnard resigned.[13]

At its March 1889 meeting, the trustees took up two proposals that prefigured the full-scale introduction of the natural sciences at Columbia. Nathaniel Lord Britton, a School of Mines PhD (1880) and longtime assistant to professor of paleontology and geology John Henry Newberry, was promoted to adjunct professor of botany. His appointment in the School of Mines, however, was objected to by Trustee William C. Schermerhorn (CC 1840), who wanted the appointment in the College of Arts, pending the university restructuring to come. The second matter was the election of the wealthy admiralty attorney businessman Charles M. Da Costa to the board. Costa's death sixteen months later and the $100,000 bequest that followed would underwrite Columbia's belated but determined entry into the natural sciences.[14]

The Da Costa gift underlined the need to settle on an organizational structure for the sciences. When the School of Mines presented its plan to take responsibility for all the sciences at the March 1891 board meeting, Schermerhorn countered by proposing a "University Faculty of Mathematics and Natural Sciences," separate from the School of Mines. A month later the trustees committed the Da Costa gift to the creation of a Department of Biology in the School of Arts. At the May meeting the trustees announced the promotion of Britton to professor of botany, not in the

School of Mines but in the School of Arts, even as they created a "University Faculty of Mathematics and Natural Sciences," to include any faculty "in any branch of mathematics or natural science who are not members of any other faculty." Point, game, match, Schermerhorn.[15]

Two months later, at the same May 4, 1891, meeting at which the trustees adopted the name of "Columbia University in the City of New York," Henry Fairfield Osborn was appointed Da Costa Professor of Zoology. Low had personally recruited Osborn from Princeton to plan, establish, and head Columbia's new Department of Biology. Joining Osborn was Edmund B. Wilson, recruited from Bryn Mawr as adjunct professor of biology, and Bashford Dean, an 1890 Columbia PhD and assistant to Newberry, as instructor of vertebrate zoology. All three were temporarily housed in the medical school until the move to Morningside in 1897, where a new building awaited, thanks to the generosity of the same William C. Schermerhorn.[16]

By then hopes for the School of Mines remaining the exclusive home of the sciences at Columbia had dimmed, although the trustees at their November 1891 meeting acknowledged that they were not ready to create "a 'School of Pure Science'… although this was held to be a probable outcome perhaps before very long." Two months later they pulled the trigger. Henceforth there would be *two* science schools. The first was to be "The School of Pure Science," initially home to the departments of biology, botany, and astronomy, but with faculty from mathematics, physics, and geology also eligible for membership. Henry Fairfield Osborn was elected dean. The second school, what had been for twenty-seven years the School of Mines, would now be "The School of Applied Science," which, at least for the time being, was home to the departments of mathematics, physics, chemistry, the four engineering specialties (mining/metallurgy, civil, electrical, mechanical), and architecture. Chandler would continue as dean. "Thus at one stroke," Low announced, "Columbia ceased to be divided into fragments, and took upon itself the aspect of a university."[17]

Neither Chandler nor his colleagues publicly challenged these developments, while some engineering faculty endorsed them. As one put it, "we would no longer have faculty who cared little about engineering." Yet, for all the administrative sense this new order made, the result was a significantly diminished role for the engineering school within the university. As if acknowledging this shift, the adjunct professor of electrical engineering Michael I. Pupin quietly shifted his principal affiliation in 1893 from Mines, which had hired him four years earlier, to the physics department in the Faculty of Pure Science. With his departure, the engineering school lost most of the services of the university's most promising and soon to be most celebrated scientist.[18]

Over the next few years, mathematics, physics, and geology all left originally assigned places in the relabeled "School of Applied Science" to take up residence in the more rarified "School of Pure Science." Architecture, thanks to the generosity of Samuel P. Avery and the lobbying of Trustee Frederick A. Schermerhorn, broke away to become in 1901 a free-standing professional school. But the coup de grace for the once pioneering engineering school came in 1902, when the trustees rescinded their 1872 authorization to the engineering faculty to bestow the PhD; henceforth the authority to grant Columbia doctorates in any of the sciences resided exclusively with the Faculty of Pure Science (where it would remain until 1959). Professional schools would later be added, degree arrangements altered, names changed, but the overarching organizational structure envisioned by Burgess and Butler, and implemented by Low in the early 1890s, was to remain in force for the next eight decades. As long as it did, the University's engineers, thought in the 1870s as about to overshadow Columbia College, were themselves consigned to the University's shadows.[19]

> And thus was this poor church left, like an ancient mother, grown older, and forsaken of her children (though not in their affections) yet in regard of their bodily presence and personal helpfulness. Her ancient members being most of them worn away by death; and these of later time being like children translated into other families, and she like a widow left only to trust in God. Thus she that had made many rich became herself poor.
>
> —William Bradford, *Of Plymouth Plantation*, on the proposed removal of the colony from Plymouth to Cape Cod, 1647[20]

THE LOW–CHANDLER DUSTUP

The relegation of the School of Mines in the reordering of Columbia University in the 1890s cannot be told wholly in terms of competing organizational plans, with some academic units winning and others losing. It has a personal dimension as well, involving arguably Columbia's best president and the engineers' most famous dean.

The question why Low, who in most official dealings was the most amiable of men, found the anything but pugnacious engineering dean not to be suffered gladly, bears scrutiny. Having attended Columbia College (Class of 1870) at a time the School of Mines seemed to some to be "overshadowing" the College, he may have had a longstanding complaint with its very existence. And while not a Knickerbocker by birth, Low may well have had personal reasons to regard the outsider Chandler and his upstart school with suspicion.[21]

Other differences put the two men at odds. Thanks to his grandfather's and father's business successes, Low was financially independent, uninvolved in moneymaking, and devoted to the unremunerated and disinterested public service expected of the city's "Best Men." By contrast, the self-made Chandler struck some old money New Yorkers as having too much of an eye for the pecuniary chance. Yet another difference between the two was that Chandler, for all his income producing pursuits, was a committed man of science, a field of intellectual endeavor that held no interest for Columbia's new president.[22]

Chandler brought some of his troubles with Low upon himself. In the matter of junior appointments to the Mines faculty, he was often cavalier about his choices. Among his favored assistants was Pierre de Peyster Ricketts, an 1870 Mines graduate who went on in 1875 to receive one of the school's first PhDs, but who also assisted Chandler in his commercial activities. Chandler regularly recommended Ricketts for promotion and salary increases, which produced grumbling among less favored junior faculty, which reached alumni, Trustee William G. Schermerhorn, and *The New York Times*. More egregious was Chandler's appointment of Charles Pellew to an instructorship months after Pellew married Chandler's daughter.[23]

Chandler regularly comingled official responsibilities with private gain. His incessant wheeling and dealing, his very ubiquity, invited suspicion. And when combined with a guilelessness in dealings with the high and mighty ("Charles was never one to hold grudges," a friend allowed), he was, after three decades as the unsupervised lord of his domain, ripe for a fall.[24]

It was Chandler's outside work that most offended Low and like-minded trustees. They could not believe such work did not detract from his effectiveness as a teacher and dean. Thus, occasional complaints, as in 1891 about a missed class, which about any other faculty member might have been ignored, in the case of Chandler, Low saw as prima facie evidence of his dean's neglect of duty.[25]

Again, Chandler's critics had a point. From his first days in New York he focused a big share of his outsized energies on becoming a player in the commercial life of his adopted city. Both Egleston, a man of independent wealth, and Vinton, a man of limited energy, occasionally took on consulting assignments, but neither on the scale or with the frequency of their enterprising young colleague. As early as 1866, Chandler hired himself out as a consultant and "industrial chemist" to New York–based corporations, which eventually included Standard Oil, the National Biscuit Company, at least two sugar refineries, and a German dyestuff company. That same year he accepted an arrangement by which he would work at Booth and Edgar Sugar Refinery on King and West Street from 6:00 to 8:00 five mornings a week (hours stipulated to allow Chandler to get back uptown to teach his 10:00 A.M. class) for an annual salary of fifteen hundred dollars. Most evenings he spent in his private laboratory set up at home, conducting tests and preparing testimony for court. He charged between one hundred and two hundred dollars a day for his consulting services and an equal amount for time as an expert witness. These activities regularly earned him an annual income twice that of his academic salary.[26]

Many of these consultancies came about through Chandler's active club life, which put him regularly in the convivial company of the city's major businessmen. He was a founding member of the University Club and, along with most of the Columbia trustees, a member of the Century Club.

But it was at the even more exclusive Farmers' Club, whose membership was limited to one hundred of the city's leading "agriculturalists," including Vanderbilts, Rockefellers, and Morgans, where Chandler alone among Columbians socialized. His substantial brownstone on East Forty-Fourth Street, equipped with a private laboratory, and after 1879 a summer home in Bridgehampton, Long Island, which entailed membership in the Shinnecock Golf Club, and regular transatlantic passages aboard luxury liners, kept Chandler in the regular company of these "farmers." Some Gotham nabobs, such as the sugar magnate Theodore Havemeyer, who provided $450,000 for a chemistry building on the new campus, became benefactors of Columbia through Chandler's solicitations. For all the good he did Columbia in directing the city's largesse Columbia's way, one can imagine it striking some trustees as operating above his job description and place.[27]

On January 5, 1890, the newly installed Seth Low commenced his presidential communications with Chandler by asking him to describe his duties as dean. Chandler responded promptly, apparently inferring no criticism in the query. The next surviving exchange has Chandler apologizing for having missed a class. That Low wrote personally on such a minor matter suggests that Chandler was under surveillance. The next letter from Low, dated November 24, 1891, informed Chandler of trustee plans to go forward with a Faculty of Pure Science as distinct from the Faculty of the School of Mines, which would be renamed the Faculty of Applied Science. Chemistry, the president further informed the senior professor of chemistry, would be represented in both faculties. "I think the University ought to be large enough to encourage the development of both kinds of study, side by side." What Low did not ask was what the university's senior scientist and longest serving dean thought about these decisions.[28]

Four months later the forty-two-year-old Low chastised the fifty-seven-year-old Chandler for submitting sloppy course catalog copy, which must have caused the blasé dean to realize that people in high places were on his case. A subsequent presidential query about his hiring practices was almost certainly intended to let Chandler know that the appointment of his son-in-law had not gone unnoted by the trustees.[29]

FIGURE 3.1 Seth Low (1815–1916), civic leader and philanthropist; oversaw the reorganization of Columbia as its eleventh president (1890–1901).

Source: Columbia University Archives.

Low's initial exchanges with Chandler took on a personal character after the president's receipt of a letter from Chandler in February 1891. After congratulating the president for Columbia's recent absorption of the College of Physicians and Surgeons, Chandler, alluding to the fact that since 1877 he had received an annual salary from P & S equal to the one he received from Columbia, sought to confirm "that I shall not suffer any pecuniary loss from the union." If there was anything about Chandler's extended professional arrangements that laid him open to trustee criticism, it was his multiple professorships and their cumulative salaries. Chandler had stepped on the third rail.[30]

At this point the archival record gets murky. One source has Low so agitated with Chandler's highhanded ways that he proposed to the trustees that he be fired, and that only the intervention of Elihu Root, the city's leading attorney, a social acquaintance of the Lows and a close friend of Chandler, kept Low from proceeding. Nothing in either the Low or Chandler papers between 1892 and 1896 confirms or refutes this story.[31]

Meanwhile, the chairman of the finance committee of the board, George Rives, conducted a review of the School of Mines. The committee's 1894 *Report* was scathing. Among its findings:

1. Enrollments in the school (excluding architecture students) had been flat for 20 years;
2. During that same period the school's costs doubled;
3. Meanwhile, enrollments among the school's competitors had grown (especially MIT);
4. The school's minimum entry age of 18 made it uncompetitive with engineering schools that had entry ages of 17 (MIT) or 16 (Cornell).

The committee's summary recommendation: "That the School of Mines be continued as a purely professional school and the provision for general scientific education be provided for in the College proper."[32]

Chandler's defensive response to this assault on his deanship only exacerbated his already strained relations with key trustees. When he cited lack of space for why Mines had only 380 enrollments, while MIT had 1,157 and Cornell 1,184, Rives dismissed it "as a confession and avoidance." Indeed, Low seems to have been sufficiently concerned about trustee reaction that at least on the issue of cost increases he sided with Chandler that the trustee *Report* may have exaggerated.[33]

The next surviving presidential correspondence with the engineering school dean has Low writing to Chandler in late 1896, when construction of the Morningside campus was well advanced. He began with a pronouncement: "When we move to the new site in the autumn of next year, a new epoch will begin in the life of the University." He then informed Chandler that his place in this "new epoch" would differ from that in the old. He could remain professor of chemistry, with its $7,500 salary, and would have an office and lab in the about-to-open Havemeyer Hall. (No great gesture this, as Chandler had personally solicited the money for the building.) Chandler could also remain dean of the Faculty of Applied Science (with a $1,000 stipend). But retaining these two positions required his resigning his separate professorship of applied chemistry in the Faculty of Applied Science ($7,500), his professorship at P & S ($7,500), and his professorship in the School of Pharmacy ($7,500 salary), which had been folded into the university a year earlier. Should he decline to resign any of these three posts, he would be obliged to resign his professorship and his deanship. In one two-page letter, Chandler had lost three of his four professorships and taken a $22,500 salary cut.[34]

The last known correspondence between them is Chandler's letter to President Low three weeks later, resigning his deanship, effective July 1, 1897. Although Low remained president for four more years and a trustee for a dozen after that, and Chandler remained on the faculty for another thirteen years, they seem to have had no occasion to correspond further.[35]

Low and trustees Schermerhorn, Rives, and Parsons had one more go at Chandler in 1898, when they endorsed the findings of an alumni committee that faulted him on the usual counts: he had not published in years; his private consulting cut into his teaching; he was a "discursive lecturer"; his most recent appointments to the chemistry department were "unjustified." But by then the once brightest star in the Columbia galaxy had the candlepower of a guttering taper and even Low seems to have decided to let up.[36]

Low's ten-year presidency ended in 1901 when the trustees declined to grant him a requested leave of absence to run for mayor of the City of New York, obliging him to resign instead. One of their concerns was that if a leave were granted, Nicholas Murray Butler, the thirty-nine-year-old dean of the Faculty of Philosophy and successor-presumptive to Low, would take one of the several university presidencies he reported being offered elsewhere. The trustees moved quickly in the spring of 1902 to appoint Butler the university's twelfth president. He would hold that position for forty-three years, eighteen years longer than any Columbia president to date.[37]

For most of those years, Butler paid little attention to his engineering school. It goes unmentioned in the index to Michael Rosenthal's estimable 528-page biography of *Nicholas Miraculous: The Amazing Career of the Redoubtable Dr. Nicholas Murray Butler* (2006). Unlike Low, however, Butler looked more kindly on Chandler's eccentricities and tolerated his outside work. When, a year into this presidency, Butler accepted a request that he host a meeting of the Society of Chemical Industry, Chandler sounded relieved: "I am very glad that you take such an interest in the matter." During Chandler's last years on the faculty, he became something of a living monument to be wheeled out at ceremonial events, such as the fortieth anniversary of the engineering school in 1904, where a commissioned portrait was unveiled and he was honored as the school's last original and longest serving faculty member.[38]

Grander still was Chandler's black-tie retirement party at the Waldorf Astoria on April 2, 1910. On this occasion, whatever disagreements he once had with some of "his boys" among the six hundred alumni in attendance, all was forgotten. Although invited to Chandler's retirement party, Low did not attend. President Butler described the seventy-four-year-old honoree as having "long been a point of contact between the University and the public, between science and industry and the public health." For his part, Chandler likely repeated what he had told Butler in an earlier retirement letter: "I have had my fair share of this most agreeable life."[39]

Having recently married a woman twenty-eight years his junior (his first wife died in 1888), Chandler lived another fifteen years in retirement, before dying in 1925. Chandler Chemistry Laboratories, construction of which began that year, bears his name. For all the problems of his later years, Chandler had for three decades provided his beloved School of Mines confident and effective leadership, even as he pioneered in the role Douglas Sloan has called "the academic as civic scientist." Whenever today's Columbia engineering graduates or faculty take up a technological challenge in the public sector, they honor Chandler's memory.[40]

"NICHOLAS MIRACULOUS" AND
HIS ENGINEERING DEANS

Chandler's place as dean was filled in 1897 by a vote of senior members of the Faculty of Applied Science. Professor of Mines Henry Smith Munroe, a graduate of the third class of School of Mines (EM 1869), one of its first PhDs (1876), and after Chandler the school's longest serving professor,

NEW YORK & JERSEY RAILROAD COMPANY
Hudson Tunnel
Inspection of the Works by Members of the Faculty and
Engineering Students of Columbia University on March 22, 1904,
after the connection of the States of New Jersey and New York
through the North Tunnel.

Chief Engineer.

FIGURE 3.2 Columbia engineering students in the Hudson Tunnel, 1904.

Source: Columbia University Archives.

was elected. Munroe served two years uneventfully before he returned to teaching and an active consulting practice. He was succeeded by Frederick R. Hutton, professor of mechanical engineering, next most senior member (1878) and another double graduate (CE 1876, PhD 1881). Butler limited his correspondence to his inherited dean to passing along complaints of parents that engineering students were "overburdened," sometimes mixed with complaints about "wholesale cribbing in examinations." Hutton's standing with the new president and the chairman of the trustee committee on education, William Barclay Parsons (CE 1882), is inferable from Butler's response to his dean's suggestion in 1903 that the engineering school develop a course on "Motor Vehicles": "Parson says 'why not a course in bicycles or naptha launches?'"[41]

In 1905, as Hutton approached his twenty-fifth year of service, Butler called upon him to retire the deanship and his professorship. Abandoning the tradition of having the engineering faculty elect its dean from among their ranks, Butler opted for a presidential appointment. Frederick A. Goetz was head of the university's Department of Building and Grounds, which he had joined twenty years earlier in an entry position. He was not an engineer or an educator or even a college graduate, having dropped out of Cooper Union after two years. He was, however, a trusted member of Butler's administrative team, so much so that throughout his deanship he remained on call as the president's "consulting engineer," using separate stationery for each role. Overseeing construction of Kent and Avery Halls, as well as replacing broken chairs in the president's dining room, fell to him. While Goetz occasionally complained, he usually went along, seeing his real job as the president's dogsbody.[42]

Goetz took as his charge as dean "to get our departments in the Schools of Mines, Engineering and Chemistry on an energetic and progressive business basis." This he set about to accomplish by finding in each of the several departments a senior member with sufficient personal authority to direct the affairs of his department without giving offense to Butler or the trustees. These "leaders" were given a virtual lifetime lock on their department chairmanships. Charles E. Lucke (ME 1904), who replaced Hutton as chair of the Department of Mechanical Engineering in 1908, was one instance, his chairmanship extending over four decades. Earl Lovell, in civil engineering, was another, holding his chairmanship for a quarter century.[43]

The new dean spent his first months in office fussing with trustees John B. Pine and William Barclay Parsons (EM 1882) about alternatives to the school's confusion of names. Older alumni continued to call it "The School of Mines," while Low's relabeling had it "The School of Applied Science." Goetz's preferred name for the currently unwieldy "Schools of Mining and Metallurgy, Civil Engineering, Mechanical Engineering, and the School of Applied Chemistry" was "The Schools of Engineering of Columbia University," which he hoped would in general usage be shortened to "Columbia Tech." Pine, the clerk of the trustees, rejected Goetz's suggestion as not allowed by University Statutes, while Parsons responded: "For my part, I have always preferred the name of 'School of Engineering' or 'College of Engineering' to 'Technology.' I prefer either of them to 'Applied Science.'" The upshot of these discussions was the school's official name became "The Schools of Mines, Engineering and Chemistry," its fourth in fifty years, and counting.[44]

Goetz was then tasked by Butler and the trustees to look into "the constant receipt of unfavorable criticisms upon the teaching" in the engineering school. In particular, reports of senior engineering faculty shifting their teaching duties onto junior members, whose manners "are said to be very bad." Butler placed the blame for "very little real teaching" that transpired in the engineering school on its tradition of giving "too free a hand to the individual departments," something he and the trustees were determined to alter.[45]

One way they intended to do so was by bringing in new blood in the person of Arthur L. Walker (PhB 1883) as Howe's successor as professor of metallurgy. The appointment was underwritten by the mining magnate and donor of the School of Mines building, Adolf Lewisohn.[46]

But as the school's administrative arrangements were tightened, its standing vis-à-vis other engineering schools deteriorated. Columbia's once premier position among mining schools, as Goetz acknowledged to Dean Frederick Keppel, had been ceded to the Colorado School of Mines, which

ENGINEERING BUILDING

FIGURE 3.3 Sketch of Engineering Hall, opened on the Morningside campus in 1898. With opening of Mudd in 1962, renamed and reassigned as Mathematics.

Source: Columbia University Archives.

was "rapidly growing in efficiency and prestige," and was attracting eastern students with its "very nominal" tuition. Closer to home, MIT and Cornell enrollments in civil and mechanical engineering grew apace, while their electrical engineering programs expanded so rapidly that by 1910 the two schools had enrollments four times those of Columbia. Unlike the physically constrained Columbia, Cornell had ample space on campus for expansion, while MIT's move from Boston to a 168-acre campus in Cambridge in 1916 and the fundraising attending the move gave it the room and means to keep growing. Perhaps as important, neither Cornell's nor MIT's engineers operated within an institutional culture, as Columbia's engineers did, that questioned their very presence.[47]

Goetz stayed on as dean for nine years. In 1916, upon the death of the university's longtime treasurer, Stephen Nash, Butler appointed Goetz to that position, in which he served faithfully through most of Butler's long presidency. As dean he advanced no major curricular proposals, undertook no new research initiatives, or seriously challenged standing arrangements. This was likely as Butler intended, because three years into his deanship, on April 13, 1909, the president informed his dean that he and the trustees believed it "time to consider the possible reconstruction of their program of studies."[48]

THE 3/3 PLAN: "IN ADVANCE OF GENERAL THOUGHT AND PRACTICE"

"Our grandfathers attempted it as an advanced or graduate school," President Butler wrote in 1931 to his fourth engineering dean, Joseph W. Barker, by way of providing a historical context to the curricular situation. "But they found themselves at least a half century in advance of general thought and practice. They had to drop their standard of admission and change their methods of work in order to deal with the facts as they then were." The advantage to this narrative was that it cleared Butler of responsibility for the decision that nearly destroyed the engineering school and laid it instead on trustees no longer living. A good idea has a plentitude of self-identified fathers; establishing the paternity of a bad one is harder.[49]

And so it was with the "3/3 plan," whereby the Columbia School of Mines, Engineering and Chemistry abandoned its practice—and that of nearly all its peers—of admitting high school graduates into a four-year bachelor's degree program in one of the constituent fields of engineering. In its place came a policy limiting recruitment to applicants with three years of prior collegiate experience at Columbia College or one of the about-to be-designated "cooperating" schools with pre-engineering programs, who would then spend another three years in the Columbia engineering school before graduating with the same bachelor's degree available at MIT, Cornell, nearby Brooklyn Polytechnic, and just about every other engineering school in the country, in four years.[50]

By the early 1920s the devastating impact of the 3/3 plan on enrollments was widely acknowledged, not least by one member of the engineering faculty who called it "near suicidal." How it came into being remains in dispute. President Butler traced its genesis to proposals emanating from the trustees, in particular two School of Mines graduates who served successively as alumni-elected trustees, Benjamin B. Lawrence (CE 1888, alumni trustee 1909–1915, life trustee 1915–1921) and Arthur Dwight (EM 1885, alumni trustee 1915–1921). Correspondence in the Goetz papers, however, suggests that Butler himself, if not the father of the plan, midwifed its delivery.[51]

Some of the thinking that went into changing the character of Columbia's engineering school can be traced back to the 1890s and President Low's insistence that "The School of Mines is more completely a professional school than any other in the country." He did so while agreeing with agitated trustees with sons in the College that the school is "not attractive to students who want a general scientific education." The trouble with Low's "professional-school" grouping was that it put engineering with Columbia's law school and medical schools, both of which in the early 1890s successfully limited admission to college graduates (or in the case of Columbia undergraduates, three years of college) and extended the curriculum from two to three years. They did so at the urging of their respected professional societies, which regarded the change as a means of simultaneously upgrading and limiting entry into what some saw as their already crowded and insufficiently exclusive callings.[52]

No such clamor for upgrading engineering education prompted Low's remarks. None of the various engineering societies advocated lengthening the curriculum of engineering programs or limiting admission to college juniors. Perhaps recognizing the problem, or heeding Chandler's objections that doing so would put Columbia's engineering school at a competitive disadvantage with all four-year engineering schools, Low declined trustee suggestions to halt the engineering school's admitting applicants directly from high school.[53]

In 1909 Butler and his trustees revisited the issue. One reason for their doing so was that again engineering enrollments exceeded those of Columbia College. To reduce the likelihood of this continuing, the trustees transferred responsibility for "the admission of school boys" from the engineering school to the College. Goetz objected but to no avail. Moreover, when Frederick P. Keppel, Butler's thirty-four-year-old secretary, was named dean in 1910, the College acquired a well-connected College man ready to advance its interests. Later that same year, when Goetz floated the idea of a new undergraduate school with "an emphasis on scientific subjects," under the auspices of the School of Mines, Engineering and Chemistry, Keppel killed it with a single letter to Butler. Such a school, Keppel wrote, was not what he understood trustees had in mind for their sons.[54]

Butler meanwhile extracted Goetz's support for—or at least acquiescence in—extending the engineering school's four-year curriculum to five. "Such a plan," he wrote in early 1910, "would hasten the day when the first year, or even the first two years, of the engineering course could be transferred to the College and the Schools of Applied Science raised to the same position as those of Law and Medicine." Goetz dutifully carried this proposal to his faculty, whom he described as positively disposed to extending the curriculum. Butler was almost instantly back to him to up the ante: "Could we not kill two birds with one stone by making the first year of the [engineering school] course identical with the first year of the College course and taking students in through the College instead of from the high schools direct?"[55]

On November 7, 1910, Butler made a third pass at Goetz, this time with a five-point comprehensive scheme "to help you strengthen the Schools of Mines, Chemistry and Engineering."

1. Engineering students to come from Columbia College or a college/scientific school "of similar standing";
2. The College to introduce a two-year pre-engineering program as prescribed by the engineering faculty but taught by College faculty;
3. A degree program of two years of pre-engineering in the College and two years in the engineering school that would yield a BS in engineering;

4. A degree program of two years of pre-engineering in the College and three years in the engineering school would yield a professional degree (ME, CE, EE, ME) in engineering;

5. The engineering faculty to "develop at the same time advanced instruction of research in engineering leading to the degree of Doctor of Engineering on a place with the degree of Doctor of Philosophy."[56]

Goetz took this plan to his three faculties and came away with the backing of two of them for extending the curriculum to five years. The opposition came from the chemistry faculty, who correctly saw the plan as likely to bring about a schism between them and the chemists in the Faculty of Philosophy. Goetz reported additional misgivings about whether handing over a share of its current teaching responsibilities to the College would lead to a reduction in the size of the engineering faculty. On the point of restoring to the engineering faculty the authority to award the doctorate, he reported unanimous support.[57]

Unfortunately, Butler had not cleared this last point with the board before transmitting it to Goetz. As for restoring the engineering faculty's authority to award the doctorate, he informed Goetz a week later, the chair of the education committee, William Barclay Parsons, voiced "strong opposition to the proposal." And as for staffing, Butler told Goetz that, "In fact, we are hoping that we may reduce our Faculty to a strictly Engineering Faculty corresponding to that of Law and Medicine." And, by the way, the trustees were now leaning toward a six-year program![58]

On December 29, 1911, the trustees announced plans to begin implementing what was now referred to as "the 3/3 plan" by accepting the first class of pre-engineering students to the College in the fall of 1913. Everyone expected the new program to start slowly, with Goetz hoping for an inaugural class of forty-five pre-engineering freshmen entering the College in the fall. When only twenty-five signed up for the 3/3 plan, the writing was on the wall. "I realize that this will be quite expensive," Goetz told Butler in his request for funds to publicize the 3/3 program in light of the disappointing numbers, "but I am positive it will not only repay us financially but it will prevent our going through the same experience as the Harvard Graduate School of

Applied Science, which has now less than seventy-five students."[59] When a Utah-based alumnus questioned the viability of the new program, Goetz acknowledged that "When we place our engineering courses on a graduate basis, . . . we shall have to depend very largely upon the technical schools throughout the country to send students to us." Why juniors at schools with one year to go for their engineering degrees would come to Columbia for three more years to earn the same degree Goetz had no clue.[60]

Goetz did not try to secure reconsideration of the trustees' plan. Other responsibilities beckoned. In the fall of 1914 he had all but assumed the duties of university treasurer, which, along with overseeing the completion of the Avery Library and Kent Hall, occupied his last months as dean. One of his final communications to Butler as dean was to report grumbling among his senior faculty about the 3/3 plan's full implementation. Having registered their concerns, Goetz characteristically demurred, assuring the president his full support "if you and the Trustees [Parsons/Lawrence] have this matter of consolidation seriously in mind." His appointment as university treasurer took effect in the spring of 1916, even as the 3/3 plan approached full implementation and the disastrous consequences were already obvious.[61]

To give it its due, the 3/3 plan had its upsides. By limiting the engineering school's recruitment pool to Columbia College juniors and to a handful of students "comparably prepared elsewhere," the plan ended the half century of competition between the engineering school and the College for high school graduates. This outcome was not lost on the protective champions of the College. Also, by eliminating the first two years of the engineering school curriculum, in effect ceding it to Columbia College, the plan eliminated the criticism of introductory engineering courses that had exercised Butler and the trustees. It also allowed Butler and Goetz to reduce the school's nonprofessorial teaching staff.[62]

Columbia College students who chose the 3/3 plan were better prepared than the typical incoming engineering student elsewhere and experienced a much lower dropout rate than at other engineering schools. At Georgia Tech or Wisconsin or Michigan, for example, attrition regularly exceeded 60 percent among first-year students. It may also have been the case that the typical Columbia engineering undergraduate was more broadly informed about the world beyond engineering than his peers at MIT, Cornell, or Stanford. The problem with the plan was simply that so few would-be engineers had two years to spare.[63]

A CONSPIRATORIAL CODA: ENGINEERS AND THE "HEBREW PROBLEM"

Another possible, albeit conspiracy-infused explanation for why Butler, some trustees, and the dean of Columbia College, Frederick Keppel, were so insistent on implementing the 3/3 plan, and later why Butler, trustees, and Keppel's successor, Herbert Hawkes, resisted junking it, suggests itself. Drastically reducing the number of engineering undergraduates may have served a larger purpose at the time: to limit the number of Jewish students admitted to Columbia. Both the timing of the plan's introduction, the half-dozen years before World War I, and those championing it, including Butler, Dean Keppel, and some trustees, but especially Benjamin B. Lawrence (EM 1878), allow the inference.[64]

Like Goetz, Keppel owed his early career prospects at Columbia entirely to Butler. As the president's assistant secretary and then secretary to the university, he had regular contact with the trustees. In 1910, when Butler appointed him dean of Columbia College, the president accompanied the appointment with a list of seventeen problems that he and the trustees wanted promptly addressed. Keppel discussed each in a memorandum written to Butler within days of his assuming the office, saving for last "the careful and sympathetic consideration of the Hebrew problem."[65]

The preoccupation of the Columbia trustees in the early 1900s with what Keppel called "the Hebrew problem" has been documented elsewhere. Hardly a trustee meeting passed without the subject being aired. The surviving contemporary correspondence among trustees and the president is largely given over to related questions:

—How might the board limit the admission of academically qualified Jewish applicants from the city's public schools, so as to make
Columbia once again the college of choice for the sons of the trustees and for alumni families of "our kind"?[66]

—What response should the board make to calls upon it to recognize
Jewish holidays or make available St. Paul's Chapel to Jewish student groups?

—Should the board consider electing a Jew to its membership? And if
so, what kind of Jew?

—Should Columbia actively solicit the financial support of wealthy
Jewish New Yorkers?

I have argued elsewhere that the timing of Keppel's job as dean of the
College, rather than any personal inclination toward anti-Semitism, made
restricting the admission of Jews to Columbia of overriding importance
during his eight-year tenure. One of the first ways he went about doing so
was to centralize admissions for the College, the engineering school, and
Barnard College in a committee that reported to him. This was done, he
informed Butler in 1911, "to ensure to the schoolmasters a continuity and
uniformity of attitude, the lack of which in their relations to Columbia they
have deplored in the past."[67]

Following a change in admissions requirements early in Butler's presidency that eliminated a classical language requirement, which had the unintended effect of making graduates of New York's public high schools qualified
applicants, Columbia experienced a rise in Jewish enrollments. Keppel never
provided any numbers and sometimes sought to minimize the magnitude
of the increase, but it would seem that on the eve of the Great War about
15 percent of the students in the College were Jewish. The same or a slightly
larger percentage likely obtained in the engineering school. "Most of these
[Hebrew students] are excellent and desirable students," Keppel informed
President Butler in 1912, "but the danger of their preponderating over the students of the older American stocks is not an imaginary one. This has already
happened at New York University and at the City College of New York."[68]

In 1913 trustees proposed a plan to require residency for Columbia College, the intent being to cut down on applications from Jewish New York City residents, who were likely to commute. When Keppel objected he did so not because the intent of the residency requirement was blatantly anti-Semitic, but because "it should fail to bring us the sons of our rich alumni, or to eliminate the Jewish element." "To put it frankly," he wrote Butler, "I do not think such a plan, or any other, would bring to us the sons of men like Mr. Rives, Mr. Cutting and Mr. Parsons." He continued in the same fatalistic vein: "If the expectation is that students of Jewish birth will no longer be attracted to Columbia College, this also is I think illusory; of the College students now living in the dormitories, a fair proportion are Jews."[69]

Keppel ended his letter by proposing an alternative strategy of capping the enrollment of entering classes at 550. Once reached, the cap would give College officials a rationale for rejecting otherwise academically eligible Jewish applicants, including those to the College's new pre-engineering program. When the Columbia College committee on instruction declared the trustees' proposed residency requirement "suicidal," the trustees settled for Keppel's capping proposal, but not before Butler instructed his dean: "I suggest treating the candidate for graduation as one treats a candidate for admission to a club, that is, having his personal qualifications examined." At the time Keppel, Butler, and nearly all the trustees belonged to the Century Association and the Columbia University Club, both of which excluded Jews.[70]

The "Hebrew problem" vexed some trustees more than others. Seth Low, who remained a trustee for a decade after resigning the presidency in 1901, favored electing a Jew to the board and left the board in 1910 when it denied use of St. Paul's Chapel to a Jewish student group. The board's lone Catholic board member, Frederick R. Coudert (AB 1890, trustee 1912–1954), seems also to have kept clear of these discussions. But others on the board—among them the board secretary, John B. Pine; the chair of the finance committee, Francis S. Bangs (AB 1878, LLB 1880, trustee 1900–1920); and the chair of the education committee (and later board chairman), William Barclay Parsons (CC 1879, CE 1882, trustee 1897–1932)—regarded the "Hebrew problem" as a matter of institutional life or death. And to their ranks was added in April 1909 Benjamin B. Lawrence (EM 1879), the first alumni-elected trustee to this heretofore self-selecting board.[71]

Lest we confuse the democratic manner of Lawrence's election to the board (he was subsequently elected by fellow trustees to a lifetime seat) with a populist disposition, here are his evaluative views on two trustee prospects:

1. If there is anything in tradition, or in antecedents, he will qualify better than any man I know in New York. He is a direct descendant of Robert Morris, and his family connection is undoubted;

2. He may have a lot of money, but he is not a gentleman by birth although he may be so by education.[72]

Lawrence wasted little time making the "Hebrew problem" his own. Eight months into his six-year term he asked Secretary Keppel for "the number of Hebrews in the different departments of the University." Keppel responded that the university had no hard numbers, but "I know that the actual proportion is considerably lower than is popularly supposed." Further correspondence on this topic has not survived, but it is at least inferable from Lawrence's subsequent lead in developing the 3/3 plan, promoting it among fellow trustees and engineering school alumni, that Keppel's response did not allay his concerns. The 3/3 plan was likely for him and others on the board in some measure responsive to the "Hebrew problem."[73]

When in 1930 Butler retrospectively dissociated himself from the 3/3 plan by declaring it the work of "our grandfathers," he fingered the then deceased Lawrence as one of its prime movers. But what of Butler himself? Here again the historical record goes thin, suspiciously so since Butler oversaw the purging of his files during the last two years of his life, when the university was threatened with law suits charging authorities with religious discrimination in admissions policies to the College and the professional schools. His refusal to reopen the question of allowing the engineering school to resume admitting students directly out of high school when both the engineering faculty and the engineering alumni association recommended doing so in the late 1920s is hardly exculpatory. Nor is his selection of Herbert Hawkes to succeed Keppel as dean of Columbia College. By his words and deeds over his quarter century as dean, Hawkes more often laid himself open to the charge of anti-Semitism than did his more circumspect predecessor. Two things for certain, Butler's insistence in 1947 that Columbia had never engaged in discriminatory policies on the grounds of religion defies belief, as does the happy thought that these policies left Columbia's engineering school unscathed.[74]

THE PEGRAM RECEIVORSHIP

Implementation of the 3/3 plan fell to Goetz's successor as dean of the Schools of Mining, Engineering, and Chemistry. To this thankless job Butler called George Braxton Pegram, a young physicist (CU PhD 1907) from North Carolina who had demonstrated an early capacity for academic administration as secretary of the physics department under William Hallock, and after Hallock's death in 1913, as department chair. Pegram taught mechanics and back in 1909 had sought assurances from President Butler "that our engineering schools shall stand for the broadest training in the fundamental sciences that underlie all engineering practice." When the president asked him to become acting dean in February 1917, "until a Dean is chosen," he reluctantly agreed, and only for a year, when "I shall welcome the opportunity of turning over the work."[75]

In April 1918 Butler informed Pegram that his appointment by the trustees as the fifth dean of the engineering school was "as inevitable as appropriate." Pegram accepted the assignment, acknowledging that "I have not looked forward to the deanship as a permanent position." He exacted the conditions that he remain chair of the physics department and a member of the Faculty of Pure Science. However reluctantly undertaken, he would be dean for thirteen years, second only to Chandler in length of service. For the second time in a row, the engineering school had at its head a nonengineer who took the job because he could not refuse.[76]

Pegram assumed his new duties as acting dean in the spring of 1917, just as the last and largest undergraduate class (143) under the old plan prepared to graduate. That fall, when Pegram reported seventy-five enrollments for the entire school under the now fully implemented 3/3 plan, he called them "very precious material." Some of the free fall in enrollments was attributed to American entry into the Great War in Europe, which had also depressed numbers in the College. But whereas the College quickly returned to pre-war enrollment levels by the fall of 1919, total engineering school enrollments rose only to 150, less than a quarter of what they had been a decade earlier. Moreover, with the engineering school reliant for future classes on students in the College's three-year pre-engineering program, and the College's admissions capped, projections for the next years were more of the same.[77]

What is striking about this resounding rejection of the 3/3 plan by the marketplace is how little Butler, Pegram, or the trustees did about it. Trustee Benjamin Lawrence began hearing complaints from engineering alumni as early as the fall of 1917. "A considerable portion of our public," he told Butler, "do not understand the recent policies of the University in regard to the Schools of Mining, Engineering and Chemistry, and are unsympathetic to them." Yet he proposed no modifications. When Butler reported Lawrence's sentiments to Pegram, the dean responded in kind, "I believe Mr. Lawrence is right" while declaring that worried alumni displayed an "ignorance, or lack of appreciation of what Columbia is attempting in Engineering." For his part, the new dean saw no need to reconsider operating the school "on the new basis of advanced work."[78]

In his first months on the job, Pegram evinced less concern about the school's plunging enrollments of men than the specter of women gaining admission. The issue arose when in the fall of 1917 a young Russian woman inquired as to the possibility of her studying engineering at the graduate level. Pegram forwarded the request to Butler. The president responded, consistent with the fact that Columbia's three graduate faculties in the post-Burgess era were now open to women, and possibly envisioning a wartime exception, that any decision "might turn on the competence of the candidate." Concerned with Butler's apparent openness to the idea, Pegram wrote back four days later that he had broached the matter with his faculty and "a considerable majority were against it." Still, Butler equivocated. "I suppose that sooner or later it will come up in some more acute form, possibly as regards chemistry and chemical engineering." The acting dean then asked the trustees to change the university statutes so that they stated clearly that "no woman shall be admitted as a candidate in the Schools of Mining, Engineering and Chemistry." Once adopted, this statute was to remain in force until 1942.[79]

Alumni disquiet about plunging enrollments only intensified until even Butler acknowledged that the 3/3 plan was to blame. "It is in advance of public need and public sentiment," he wrote to Pegram in 1922, "and that the Schools themselves and the University are suffering in consequence." "Without surrendering the present six-year course," the president insisted as he took a first step down the slippery slope that the engineering school should consider "another and a less difficult course."[80]

Just such an alternative course was already under discussion among younger engineering faculty properly concerned with their long-term job prospects. The most deeply involved was adjunct professor of civil engineering, James Kip Finch. In keeping with Butler's charge to Pegram, the Finch-led faculty goal devised a shorter and less academically demanding alternative to the 3/3 plan. This would be accomplished in two steps: reduce admission requirements to match those of the College, and reduce the number of courses required before transferring from the College to the engineering school, thus enabling transfer after two years. This effectively reduced the program leading to a professional degree from the currently prescribed six years (3/3) to five (3/2). To make the current six-year program still more attractive, its completion would now be recognized by both an AB and a professional degree. Finally, after rejecting the faculty's call for restoring the four-year program leading to professional degrees (the norm elsewhere), the trustees agreed to provisions by which any student in the College could transfer after two years into the engineering school, and where, after two years there, could graduate with a generic BS degree (2/2).[81]

This came some way to bringing one of the engineering school's undergraduate degree programs back in line with those of the competition. The implementation of the faculty-modified arrangements in 1924 was followed by a modest uptick in pre-engineering enrollments and in transfers from the College to the engineering school. Still, writing twenty-five years later, in 1949, Finch, who more than any other person deserves credit for keeping the school from having, as he put it, "almost ended fatally," lamented that "we have not yet lived down the six-year reputation.... [It] has been a handicap we have to face even today." In point of fact, most of what little growth in enrollments the engineering school experienced during the interwar period occurred at the graduate level, and much of it from an on-again, off-again agreement with the Navy by which the school provided instruction in electrical and mechanical engineering for upwards of forty career officers annually. Several dozen naval flag officers during World War II were Columbia graduates, the best known being Captain (and later Admiral) Hyman Rickover (MS EE 1929).[82]

In 1927, a confidential *Report of the Alumni Committee on the Engineering Schools,* chaired by Milton L. Cornwell (CE 1905), made clear the distress felt among the school's most active alumni with its present condition. Among the *Report's* seventy-one findings, the bluntest related to enrollments:

> 14. There has been no appreciable growth in registration since the longer course was instituted. There are about one third as many students as we had in the old and shorter course. In the meantime, the schools around us, which we have studied, with other conditions similar to our own presumably, have more than doubled in size during the same period. We may then assume that our registration is about one-sixth of what we might have expected with a continuation and modernization of the old course.[85]
>
> 47. The longer course, without adequate alternatives, has been the chief deterrent to matriculation in our schools.[83]

The *Report* went on to discuss the financial and reputational implications of operating a school at one-sixth capacity. In a survey of recent graduates and queries to various engineering societies, no evidence was found of six-year engineers enjoying any financial or promotional advantages over their four-year counterparts at MIT or Cornell. The committee made short-shrift of the purported parallelism between engineers and medicine or law, concluding that "we are unable to say why the Engineering profession in general does not seem to show more professional jealousy to restrict entrance to the better qualified applicants." But that it did not was clear. Finally, the *Report* dismissed the initial prospect of attracting students from other colleges as totally illusionary.[84]

Among its recommendations, the *Report*, anticipating trustee opposition, came up just short of calling for an end to the requirement that high school graduates seeking engineering training at Columbia first go through two or three years of preparatory studies in Columbia College. As to the overall sentiment of alumni about the effect of the 3/3 plan on alumni support of the engineering school, it could hardly have been more pointed:

Many of our older graduates are saddened by, and some resentful of the shrinkage in the size of the schools. It is a general source of chagrin to alumni to have their Alma Mater seem to decrease in importance and prestige. Our smaller size has given this impression quite generally. We deplore the condition, not only for reasons of sentiment but also because alumni should be the natural sponsors and supporters of their University's prestige.[85]

Little came of the alumni committee's heroic efforts to restore Columbia engineering to its earlier place of prominence among the country's leading engineering schools. The trustees did, however, adopt the cosmetic recommendation that the cumbersome name "Schools of Mining, Engineering and Chemistry," be changed to "The School of Engineering." And so it became, and so it remained for the next thirty-four years, until lengthened in 1961 to "The School of Engineering and Applied Science."

Again, to be fair to Butler, little occurred at Columbia in his prime without his approval, so the initiative for the Cornell Committee may well have come from him. This could be inferred from his remarks in a confidential report delivered to trustees on May 3, 1926, a year before the release of the *Alumni Report*. The occasion was the twenty-fifth anniversary of his presidency, the subject a generally upbeat and self-congratulatory school-by-school review of the university's competitive standing. Nine of the fifteen departments in Columbia's three graduate faculties Butler adjudged to be the best in the United States, ranking the other six as "second tier," requiring "only a moderate amount of new support to give them the distinction that we wish them to have." "None better" was his summary assessment of Columbia College; "in its best condition ever," he said of the law school.[86]

Butler went on to designate Columbia's three other professional schools—medicine, journalism, and engineering—as "where the most strengthening [was] needed." The problem with engineering, as the president saw it, was that the school, at the urging of successive engineering alumni trustees (he cited Lawrence and Dwight, both dead), "went too quickly in making the course too rigorous," which produced a "sharp drop in enrollments." He paired this acknowledgment with the further admission that the school currently "lacks distinction in several departments." The engineering school had finally made it to Butler's to-do list, right behind the medical school, whose reputational problems were being addressed by plans well underway for a six-million-dollar medical center. Engineers next.[87]

Another hint that Butler had begun to take seriously the situation in engineering was the praise he lavished on specific members of his administration, among them the deans of the College, the law school, the medical school, and even the extension school; Dean Pegram of engineering went unmentioned. This may have been Butler's way of acknowledging complaints heard a year earlier from engineer alumni on and off the board that what the engineering school needed at its head was an engineer. When Butler first apprised Pegram of these grumblings in 1925, neither made much of them. But when the issue resurfaced in 1928, Pegram welcomed his deprecatory characterization as "a man of science" and took the occasion to inform his president of his desire to return full-time to the physics department. Butler's receptiveness to this request likely reflected his sense that the time to address the problems of the engineering school had come.[88]

Pegram's critics included one of the engineering school's most prominent alumni, Gano Dunn (EE 1891), a prize student of Michael Pupin, later an electrical engineer of Crocker-Wheeler Company (1891–1911) and longtime president (1911–1951) of the J. G. White Engineering Corporation. Dunn served on the 1927–1928 engineering school's alumni committee and in 1928 was elected alumni trustee. By the time Pegram informed Butler of his wish to return to the physics department, Dunn was already seeking out a successor. When Butler asked Dunn his opinion of Pegram, he replied: "[I] concur in much of the temperate criticism of the leading alumni and others, which see him lacking in magnetism and constructive leadership in the executive work of the Engineering Schools." Dunn then proposed as Pegram's successor the thirty-four-year-old Joseph W. Barker, an MIT graduate at the time finishing his first year as the chairman of the electrical engineering department at Lehigh University. When board chairman Parsons, head of the construction firm Parsons Brinckerhoff and every bit Gunn's equal as an engineering eminence, questioned the extent of Barker's administrative experience, Dunn pressed his case: "The conditions at Columbia are difficult in the extreme, and he may fail, but stagnation is never a solution."[89]

There the situation stood in the summer of 1929, three months before the Stock Market Crash in October, after which whatever heady plans Butler and his trustees had for their "stagnant" engineering school would have to wait upon prosperity's return.

"SEPARATION OF THE CHEMICAL PRECIPITATES"

This chapter closes with an incident that occurred in 1915, which sheds lights on the progressive and personal alienation of Columbia's engineers and applied scientists from their colleagues in the "pure" sciences, which began with the creation of two science faculties in 1891.

The surviving historical record provides few expressions of concern about this breach, even as it widened, and no evidence of concerted efforts to rectify it. One notable exception was Robert S. Woodward, professor of physics and the second dean of the Faculty of Pure Science. In 1904 Woodward called for the merger of the Faculties of Pure Science and Applied

Sciences into a single Faculty of Science. He called the distinction between the "pure" and "applied" sciences "unjust and invidious" to the university's ongoing efforts in the applied sciences, while unfairly privileging work in the "pure" sciences. Perhaps the fact that Woodward received his undergraduate degree in engineering at the University of Michigan in 1872 before advanced training in physics at Gottingen made him less Manichean than his Columbia colleagues. Unfortunately for Columbia's engineers, Woodward's comments came on the eve of his departure from Columbia to Washington to become president of the Carnegie Institution.[90]

One of Chandler's last acts as chairman of the chemistry department was to launch in 1906 an undergraduate program in what he called engineering chemistry. In doing so, he allowed Columbia to share bragging rights with MIT as having initiated undergraduate instruction in what later came to be called chemical engineering. Chandler then secured the professorial appointment of one of his long time assistants, Arthur Whitaker, to oversee the program. As with Chandler, Whitaker and two subsequent associates, Lloyd Metzger and Samuel Tucker, were housed in Havemeyer Hall, the home of the chemistry department, not in Engineering Hall or the School of Mines building with the other engineering faculty. Relations between the engineering chemists and the department's other chemists remained amicable so long as Chandler was chairman, but with his retirement in 1910 and the recruitment of Alexander Smith as chair of the chemistry department, the situation changed.[91]

Three things to know about Alexander Smith: First, he proudly declared himself a "pure scientist," with all the implied invidiousness that Woodward had decried. Second, he came to Columbia from the University of Chicago and was unbeholden to Chandler or his legacy. And third, during his nine-year chairmanship (1911–1919), he became increasingly mentally unbalanced and eventually committed suicide in 1922.

As chairman, according to an account written three decades later by Arthur W. Hixson, then an instructor in the department, Smith became increasingly agitated about the growth of the chemical engineering undergraduate program and space demands it put on his department. Smith believed that chemical engineering, a field now much in demand with

America's previous reliance upon Germany for chemical products disrupted by the Great War, threatened to crowd out "real" chemistry instruction and chemists of the "pure science" persuasion. In 1915 he decided to do something about it.

Sometime that spring Smith called in Whitaker for a showdown. Hixson says Smith told the equally excitable Whitaker of his "firm determination to make the Columbia Department of Chemistry the foremost school of inorganic and physical chemistry in America, even if he had to subordinate the higher branches of chemistry including chemical engineering to do it." Smith went on: "There was no room in the department for anything but pure chemistry." As for the enrollment success of chemical engineering, Smith complained that "this cow-bird engineering fledgling is taking so much space in the chemistry department's nest there will soon be no room for anything else."[92]

Whittaker promptly protested to Dean Goetz that Smith intended to evict him and his chemical engineering colleagues from Havemeyer Hall. Goetz took the problem to Butler. On June 7, 1915, the Columbia trustees effected the permanent separation of the chemistry department, which would be under the Faculty of Pure Science, while establishing a new department, chemical engineering, in the Schools of Mining, Engineering and Chemistry. They were to be housed on separate floors of Havemeyer. Whitaker left Columbia in 1917 for wartime service and later for a position in the chemical industry, as did Metzger and Tucker. Smith resigned as chair of chemistry in 1919, but not before he had effected what Hixson had a Havemeyer wag calling "the separation of the chemical precipitates."[93]

Both departments maintained high standing among their interwar peers for their teaching and research, but with virtually no interdepartmental collaboration. Eighty-two years would pass before the Columbia Department of Chemistry and the Columbia Department of Chemical Engineering affected a rapprochement and reengaged in mutually beneficial ways, about which more in Chapter 9.

4

THE GREAT DEPRESSION AND
THE GOOD WAR 1930–1945

In spite of what alumni and faculty members say,
our Engineering School still bears a high reputation.

—PROFESSOR OF MINING THOMAS READ TO DEAN JOSEPH BARKER, OCTOBER 21, 1932[1]

EE—The happiest family in our group. . . . It lacks distinction since
the deaths of Pupin and Morecroft, but presents no difficult problems.

—DEAN JOSEPH BARKER TO PRESIDENT NICHOLAS MURRAY BUTLER, OCTOBER 30, 1935[2]

T WO MOMENTOUS GLOBAL events dominate this chapter: the Great Depression and World War II. In terms of higher education, it was a period of general retrenchment in the face of financial stringencies, stagnant enrollments, and wartime dislocations. For the engineering profession, the hallmarks were uncertain job prospects and a shift in the sponsorship of large projects from the private corporate sector to the federal government. For Columbia, these years necessitated shelving plans based on assumed continued prosperity. A protracted presidential endgame followed. To the School of Engineering fell a leadership vacuum, curricular stasis, and the disruptions of war—hard times. But before that, a literary excursus.

MARTHAS AMONG THE MARYS

By the 1920s Columbians had shed their earlier skepticism about the central place of science in the university. This despite the fact that with the defection of the geneticist Thomas Hunt Morgan to the new California Institute of Technology in 1922 and of epidemiologist Hans Zinnser to Harvard the following year, its once preeminent place in the natural sciences was not ensured. Still, psychology remained top-ranked throughout the interwar period, while physics experienced a flowering upon I. I. Rabi's return from European studies in 1931. Chemistry and Astronomy also had nationally recognized faculty in Harold S. Urey and Wallace Eckert. If not as large as the Faculties of Political Science and Philosophy, Columbia's Faculty of Pure Science saw itself as every bit their equal—and was so regarded by its nonscience colleagues. Indeed, what members of all three of Columbia's graduate faculties shared was a view of the professional schools, and most particularly their School of Engineering, as having at best a marginal and derivative place in the university.

The resultant institutional ordering resembles the sibling arrangements of Martha and Mary, Lazarus's two sisters, in the *Gospel of Luke:*

> As Jesus and his disciples were on their way, he came to a village where a woman named Martha opened her home to him. She had a sister called Mary, who sat at the Lord's feet listening to what he said. But Martha was distracted by all the preparations that had to be made. She came to him and asked, "Lord, don't you care that my sister has left me to do the work by myself? Tell her to help me!"
>
> *"Martha, Martha," the Lord answered, "you are worried and upset about many things, but few things are needed—or indeed only one. Mary has chosen what is better and it will not be taken away from her."*

—THE GOSPEL OF LUKE, 10:38–42[3]

Luke's Christ reproves Martha for spending time worrying about "the many things," while crediting Mary for choosing the "better thing," in this instance, hanging worshipfully on the guest's every word while leaving to Martha preparations in the kitchen. For our purposes, Mary is the privileged scientist, whose imperative is to know, to find out, to ask the big questions, to await enlightenment; Martha is the engineer, left to the chores of hospitality, to bake the bread, to "do the work." And just as there is no question in Luke's telling whom the honored guest identifies as the sister with the higher calling, there is none between the relative standing on technology-wary campuses accorded the parlor-dwelling scientist and the engineer-in-the-kitchen. The "ascribed primacy of science," writes the historian of science and technology Paul Forman, has been "a general and consistent feature of American technological ideology from the Gilded Age into the first decades of the Cold War."[4]

Another literary text bearing on this posited relationship between science and engineering is the poetic retelling of Luke's Gospel story in Rudyard Kipling's "The Sons of Martha" (1907). The poem begins seemingly accepting of Luke's rendering and Christ's judgment:

The Sons of Mary seldom bother, for they have inherited that good part;
But the Sons of Martha favour their Mother of the careful soul and the
    troubled heart.
And because she lost her temper once, and because she was rude to the
    Lord her Guest,
Her Sons must wait upon Mary's Sons, world without end, reprieve,
    or rest.

But once Mary's progeny have their favored place acknowledged, Kipling launches a spirited case in the second stanza for the selfless self-sacrifice that is the lot of Martha's sons:

It is their care in all the ages to take the buffet and cushion the shock.

It is their care that the gear engages; it is their care that the switches lock.

It is their care that the wheels run truly; it is their care to embark and
    entrain,

Tally, transport, and deliver duly the Sons of Mary by land and main.

Four stanzas follow, all making the case for Martha's sons:

They do not preach that their God will rouse them a little before the nuts
work loose.

They do not preach that His Pity allows them to drop their job when they
damn-well choose.

Only in the last stanza does Kipling reintroduce the Sons of Mary, still
allotted "the better part":

And the Sons of Mary smile and are blessèd—they know the Angels are
on their side.

They know in them is the Grace confessèd, and for them are the Mercies
multiplied.

They sit at the feet—they hear the Word—they see how truly the Promise
runs.

They have cast their burden upon the Lord, and—the Lord He lays it on
Martha's Sons![5]

Kipling confers upon the sons of Martha a social functionality with its
own appeal, especially among those whose jobs are "to take the buffet and
cushion the shock." Not surprisingly, as the historian of technology Ruth
Oldenziel writes, "Kipling's poem struck a chord among engineers from
its publication in 1907 through the 1950s." In identifying with Kipling's
Martha, unlike Luke's Martha, who at least lodged a fair-employment
complaint with the authorities, engineers assumed, in Forman's phrase,
"the posture of proud subservience."

Forman draws on several Columbians to buttress his thesis of the
primacy of science as a characteristic of the modern era. One is John
Dewey, Columbia's most distinguished philosopher of the first half of
the twentieth century. Writing in the early 1930s on the discoveries in
electricity and radio, Dewey makes a sharp distinction between the scientist and those engaged with "a technology of apparatus, materials and
numbers." In doing so, he conflates the businessman and the engineer,
all the more to make his point:

Did the businessmen who operate our economic system produce such things? No, I say again. They were produced and invented by the scientists who were working disinterestedly and honestly and who then placed the fruits of their labor at the disposal of others . . . the pecuniary aims which have decided the social results of the use of these technologies have not flowed from the inherent nature of science.[6]

That such subservience held sway among interwar Columbia engineers can at least be inferred by the apparent ease with which they accepted an iteration of it from one of their own, Gano Dunn (1891). Dunn was a Barnard College and Columbia University trustee, president of Cooper Union, head of one of America's most prominent engineering firms, one-time president of the American Institute of Electrical Engineers, and main speaker at the Columbia School of Engineering alumni banquet in 1930, where he contrasted the scientists whose names and discoveries adorned the new National Academy of Sciences building with himself and his audience:

> They are no idle boasts, those legends written under the dome of the beautiful temple of science in Washington, 'Pilot of Industry,' 'Conqueror of Disease,' 'Multiplier of the Harvest,' 'Explorer of the Universe,' 'Revealer of Nature's Laws,' 'Eternal Guide to Truth.' The priests who sacrifice in her temple know the joys of the freedom of the human intellect . . . the scientist's ecstasy of thought.
>
> But beside her priests science has her worshippers who go out into the world and are a part of it. They are the engineers.

Accounts of the banquet suggest Dunn's remarks were well received by the engineers in his audience, with none showing any semblance of Martha's public irritation with her relegation to the kitchen.[7]

> And because she lost her temper once, and because she was rude to the
>   Lord her Guest,
> Her Sons must wait upon Mary's Sons, world without end, reprieve, or rest.

THE BARKER DEANSHIP

The movement to replace Dean Pegram with a "real engineer" had been launched at the crest of a decade of national prosperity that lavished riches upon private universities. Although Columbia had been reluctant to appeal directly to its alumni for the kinds of support that by 1920 had allowed Harvard, Princeton, and Yale to displace it as the richest university in America, its Manhattan real estate holdings seemed to ensure a financially secure future. Moreover, President Butler's 1926 report to the trustees signaled a willingness to reconsider an earlier decision to reject the services of the professional fundraising company, Price-Jones. The president also told his trustees he had personal knowledge of several wills that when probated would add to the university's coffers "not less than fifty or sixty million dollars."[8]

The Stock Market Crash of October 1929 did not immediately oblige Butler to downgrade Columbia's finances. Most of the university's endowment was in real estate, in long-term lease agreements and mortgages, such as that on the land on which Rockefeller Center was being built. Only a prolonged depression would affect these assets or the expected income from promised bequests. When recruiting a new dean for his engineering school in the summer of 1930, Butler likely believed the university's resources equal to his promise that what had been done for the medical school in the late 1920s, the building of a $20 million medical center, would now be done for the engineering school.[9]

The pledge was made to Joseph W. Barker, a thirty-eight-year-old MIT-trained electrical engineer in his first year as chair of the electrical engineering department at Lehigh University. He had been identified by a search committee headed by trustee Frederick Coykendall but personally recruited by Gano Dunn. With assurances from the search committee, the president, and the board that material support was at hand for "again placing our School in its former pre-eminent position," Barker accepted the position and took up his duties as Columbia's sixth dean of engineering in September 1930.[10]

A year later Barker informed Butler of an offer from MIT's incoming president, Karl Compton, which he was planning to accept. He cited as his reasons the engineering school's lack of space and equipment and "the constant friction and internal bickering" in several of its departments. Butler again gave assurances that the board "will do for engineering what it had been done for medicine . . . as soon as the present economic clouds lift." Engineering alumni Gano Dunn, Milton Cornwell, H. H. Porter, and Walter H. Aldridge all weighed in with promises of a forthcoming engineering center. Now two years into the Depression and the university's finances already severely straitened, such assurances were clearly wishful thinking. Nonetheless, Barker agreed to stay at Columbia, where whatever prospects he had "of again placing our School in its former pre-eminent position" remained hostage to the persistent economic realities of the Depression, an aging but entrenched faculty, and the fitful attentions of an increasingly enfeebled president.[11]

While all American universities felt the negative impact of the Great Depression, some used the decade to advance their institutions' competitive positions by strengthening or developing specific areas of coverage. Harvard is a case in point, where President James Bryant Conant, himself a Nobel chemist, restored Harvard's science departments to an eminence ceded by his predecessor. He also put in place a rigorously meritocratic tenure process. Similarly, MIT President Karl Compton and his vice president, Vannevar Bush, used the 1930s to reposition MIT, particularly its electrical engineering department, away from reliance on industrial relationships that dominated in the 1920s and toward a new focus on basic science and government contracts. Stanford is another instance, where department chairs Frederick E. Terman and Stephen P. Timoshenko transformed the electrical and aeronautical engineering departments into "steeples of excellence," thereby establishing the previously sleepy Stanford as an up-and-coming technology-embracing university. The University of California, Berkeley, made good use of the economic crisis through a combination of presidential leadership from Robert G. Sproul and faculty leadership from physicist Ernest O. Lawrence, solidifying Berkeley's standing as the country's premier public university in the sciences and engineering.[12]

Other engineering programs struggled during the Depression. Cornell's enrollments dropped by half, and at least one of its departments, electrical engineering, once preeminent, was no longer even distinguished. "The fact had to be faced," Cornell historian Morris Bishop wrote, "that Cornell was losing, even in competition with other engineering schools."[13]

The situation was worse at Columbia, where the slide in enrollments and reputational decline antedated the 1930s. Unlike Harvard, MIT, and Berkeley, Depression-era Columbia did not benefit from new, resourceful, and imaginative presidential leadership. Butler, already in his late sixties and three decades on the job at the outset of the Depression, stayed on another fourteen years, despite bouts of ill health, deafness and, at the end, partial blindness. By the late 1930s what had been earlier in Butler's presidency regular bursts of executive initiative was replaced by an elaborate presidential endgame, not over until 1945 when the trustees forced his retirement. He died two years later.[14]

Nor was there on Morningside the administrative equivalent of a Vannevar Bush, Frederick E. Terman, or Ernest O. Lawrence. Barker stayed on until 1940, when he decamped to Washington to the Navy Department, leaving the school to be administered throughout the war by acting dean James Kip Finch. Thus, Columbia and the engineering school passed through both the Depression and the war treading water. A double crisis wasted.[15]

"LITTLE PRINCIPALITIES AND POWERS": THE INTERWAR ENGINEERING FACULTY

When Barker complained in 1931 about the "constant friction and internal bickering" that characterized departmental relations, Butler had his full sympathy. The year before he had faulted senior engineering faculty for "setting up little principalities and powers which become sensitive about their own dignities and rights."[16]

Inbreeding contributed to the engineering school's insularity. The interwar faculty experienced little growth, hovering for most of those two decades in the narrow range of thirty-five to forty regular members. Access to most of these positions came from below and for the most part was limited to graduates of the school, insiders who had impressed their professors as students and their senior colleagues by a willingness to endure protracted apprenticeships before being considered for a permanent position. Civil engineering professor James Kip Finch is a case in point. After graduating with a degree in civil engineering in 1906, and working a year in industry, he taught briefly at Lafayette. Invited back to Columbia in 1910, Finch taught as an instructor while completing his MS in civil engineering two years later. He was kept on as an instructor for two additional years, during which time he took on the administration of Camp Columbia, in Litchfield, Connecticut, where incoming students received instruction in surveying. Earl Lovell, who had been the executive officer of the Civil Engineering Department since 1906 and would continue to be so until his retirement in 1934, secured Finch's appointment as assistant professor in 1914.[17]

For the next eighteen years Finch performed most of the functions of department chair, while Lovell looked after what Barker characterized as his "extensive outside interests." A competent teacher and able day-to-day administrator, Finch did little research or publishing in his field of civil engineering, choosing instead to devote his spare time to the history of engineering. As Lovell's retirement approached in 1935 and the chairmanship fell vacant, Butler gave serious thought to passing Finch over for someone less "dominated and suppressed by a petulant senior." It was Barker who urged Finch's appointment on the grounds of administrative continuity, while conceding "he will not reach great distinction for himself."[18]

Such academic indentureships were not unique to engineering or confined to the interwar period, but they were crucial to the workings of the interwar engineering school because of the high incidence of senior faculty with "extensive outside interests." Junior colleagues frequently found themselves the "workhorses" of their department chairs, who were then free to accept consulting assignments off campus. The Depression made such industry consultancies scarcer, which only made senior professors more protective of those they had.[19]

Again, outside consulting was not unique to Columbia engineers. MIT had institutionalized such faculty–industry arrangements in 1919 with its "Tech Plan," whereby large corporations like Eastman Kodak and DuPont paid annual retainers to MIT to assure them favored access to faculty. This differed from the Columbia situation in that MIT had a financial stake in these arrangements, whereas the faculty–industry hookups at Columbia had faculty operating as independent contractors, with nothing in it for Columbia or the engineering school. Not surprisingly, Columbia trustees, presidents, and even engineering deans saw these arrangements as subject to abuse.[20]

Barker's introduction to these arrangements occurred four months into his deanship. It involved Charles M. Lucke, the once bright young man whom Dean Goetz had installed as chairman of mechanical engineering back in 1908. Twenty-three years later Lucke was still chairman and everyone in the department owed his appointment to him. In early 1931 this twenty-nine-year veteran of the faculty wrote to inform his new young new dean that "I have contracted to give three days a week of my time to Babcock and Wilcox Co. as consulting engineer." He further announced that he would be away from campus Tuesdays, Wednesdays, and Fridays, and would resign his position on the Committee on Instruction, which met at noon on Fridays. He did not suggest resigning as department chair or taking a cut in salary.[21]

Barker responded with sarcasm: "I am very sorry that the date and hours set for the COI does not meet your convenience." He then cited University Statutes 60 (that went back to the Newberry incident in 1882) and 65 (as relates to "'Unofficial Employment") as in conflict with Lucke's contractual arrangements with Babcock and Wilcox. Apparently hearing nothing back, Barker pressed in a follow-up letter: "Do you intend to continue such contractual relations?" To this Lucke promptly responded in a letter containing four points:

1. That Lucke had been a member of the engineering faculty since 1902;
2. That he had been chair of his department for 23 years;
3. That he had been consulting ever since his appointment in 1902;
4. That he would be taking up the matter with President Butler.[22]

Subsequent correspondence as to the upshot of this confrontation is missing. Two weeks after Lucke's letter, however, President Butler directed Barker to have Statute 65 read to the engineering faculty at its next meeting, allowing the inference that Lucke's effort to go over Barker's head failed. Yet three years later Barker wrote to the president complaining about how Lucke "dominates his staff . . . is distracted by work for Babcock and Wilcox . . . and is not cooperative with his dean." In casting about the department for a possible replacement for Lucke as chair, Barker described one of the department's two other senior members as having "family wealth [and who] never exerted himself," the other, "not a man of distinction." Among his many talents, Lucke had apparently followed the first rule of academic survival: avoid hiring colleagues capable of replacing you.[23]

In the letter to President Butler bemoaning the staffing situation in mechanical engineering, Barker tried to find something positive to say about other faculty colleagues. The best he could manage was to declare the electrical engineering department "the happiest family of any of our group." "It lacks distinction since the deaths of Pupin and Morecroft," he admitted, before bestowing on it a dean's highest compliment, "but [it] presents no difficult problems."[24]

All of Columbia's interwar engineering departments were small as compared to their peers. In 1934 the electrical engineering department was the largest, with eight faculty. MIT had an electrical engineering department of thirty-five faculty; Lehigh had twenty. To be sure, Columbia's electrical engineers included Edwin H. Armstrong (1913), the inventor of the regenerative circuit and low-noise FM radio. Armstrong joined the department in 1934 as honorary professor and remained until he ended his own life in 1954. But Armstrong did no teaching during these two decades on the faculty and played only a minor role in departmental affairs. Much of his time when not working on his inventions in the Marcellus Dodge Laboratory in the Philosophy Hall basement was given over to bitter patent litigation battles he waged with RCA's president David Sarnoff.[25]

Barker used the same problem-laden/problem-free criterion in accessing his other departments. The six-member Department of Mines, Mineralogy and Metallurgy, which sometimes acted as three separate departments,

FIGURE 4.1 Edwin H. Armstrong (1890–1954) (EE 1914), inventor of low-noise FM radio, honorary professor of electrical engineering (1934–1954).

Source: Columbia University Archives.

and had the fewest students, he recognized for its active research agenda, at least until its most distinguished member, the metallurgist William S. Campbell, was slowed by age. He died in 1936. Barker regarded longtime department chairman and professor of mining Thomas Read as "scholarly but lacks drive . . . no mentoring skills." The younger Eric Jette and Phillip Bucky, both hired in 1929, were also scholarly enough, "but not of the administrative type." As for the department's view of itself, Read undoubtedly included his own department in his 1932 all-school benediction: "In spite of what alumni and faculty members say, our Engineering School still bears a high reputation." Three years later, Barker summed up his view of the department's three most senior members by urging that they "be retired as soon as eligible.[26]

The seven-member Chemical Engineering Department had larger interwar registrations than other departments but was, in Barker's 1935 view, "rapidly losing distinction." Again the problem was attributed to advancing age and off-campus distractions. Aside from Leo H. Baekeland, a Belgian émigré and inventor of Bakelite, who held an honorary professorship in the department since 1917, only Colin G. Fink, an electrochemist who had joined the department in 1920, acquired a national reputation for his introduction of tungsten for incandescent lamp filaments. The department lacked effective leadership. Daniel D. Jackson, a member of the department since its separation from chemistry in 1915, served as chair for twenty years. William D. Turner, an assistant professor hired in 1929, was, by Barker's description, "Jackson's workhorse." The other senior member, Ralph McKee, having been implicated in the promotion of fraudulent mining stocks in 1925, was ineligible for the chairmanship. Barker's solution to problems of the department: the forced retirement of McKee and Jackson.[27]

"The civil engineering department in particular," Barker wrote Butler in the spring of 1932, "has given me much worry." Here, again, the problem was an entrenched and, in the person of Earl Lovell, "lackadaisical" chairman, with no one in the wings. When Lovell retired in 1934, the chairmanship went by default to Finch. But unlike every other department, civil engineering had a relatively young and administratively able member in Jewell M. Garrelts, a Midwesterner who joined the department in 1927, and Donald Burmister, who came two years later; both would see the department through the war and well into the postwar period.[28]

The problem with the three-member industrial engineering department was politics, specifically the politics of its chair, Walter Rautenstrauch. A member of the mechanical engineering department since 1904, Rautenstrauch persuaded Butler in 1919 that the kinds of industry time-study/efficiency experiments identified with Frederick W. Taylor, Thorstein Veblen, and the Progressive movement merited departmental status at Columbia. The ensuing Department of Industrial Engineering was arguably the first of its kind and attracted a sizable student clientele. But by the early 1930s, when the industrial system which engineers like Rautenstrauch had been prepared to serve in the prosperous 1920s was in shambles, some now recommended drastic interventions. This movement acquired a name—"Technocracy"—and to conservatives like Butler, Barker, and Columbia trustees, a "troubling" reputation.[29]

Rautenstrauch's advocacy of public control of industry and the use of his office to conduct an energy survey of North America briefly earned Columbia the reputation in the conservative press as "Technocracy Central," which Barker regarded as an "embarrassment." Butler disavowed the movement and broke off personal relations with Rautenstrauch, while Barker, a Willkie Republican, dismissed his colleague as "a crusader [with] a martyr complex." Read today, neither his more provocatively titled *Who Gets the Money? A Study in the Economics of Scarcity and Plenty* (1939) nor his textbook, *The Design of Manufacturing Enterprises* (1941), would seem to have merited the acrimony directed his way during his last years at Columbia. "The School will breathe more freely," acting dean James Kip Finch harrumphed on the eve of Rautenstrauch's retirement in 1943, "when our curious friend, an irresponsible radical and exponent of radical business administration, is no longer connected with our staff." Students remember Rautenstrauch as a caring and insightful teacher.[30,31]

Part of Barker's problem, the Depression aside, was that he lacked the power to intervene in departmental internal affairs without support of the department chairs. Until 1937 the chairs dealt directly with Butler in budget negotiations. The dean's role was merely advisory. After 1937, when Butler's failing health obliged him to delegate authority to others, department chairs dealt with Provost Frank Fackenthal, who was perfectly at ease in acknowledging that "I know nothing about engineering."[32]

What the engineering school needed, if it were "to secure many worthwhile changes," Barker regularly informed Butler, was "new blood in nearly every department." But lacking funds and space, nothing happened. For the first five years of his deanship, Barker made no hires and saw no faculty hired away. The second five years found him regularly updating faculty lists with the earliest possible date of retirement. But with the dim job prospects that persisted through the 1930s, few faculty were likely to forego a regular—albeit frozen—salary. When the rare opening occurred in the professorial ranks, it was deemed unconscionable that the school would look outside for a new man when there were several no-longer-so-young Columbia-trained instructors and adjuncts who had for years been waiting patiently for an opening.[33]

The demographic upshot was a dramatic aging of the engineering faculty. Between 1929 and 1939 the mean age of the engineering faculty went from 50 to 59, the median age from 50 to 56, the youngest member from 29 to 42, the percentage of faculty under 40 from 20 percent to 0. From the perspective of the individual faculty member, life as an engineering professor in the 1930s had its upside. Enrollments in the 1930s fluctuated between 230 and 290, with a third of them graduate students, most of them part-time. Estimating the full-time-enrollment equivalent (FTE) at roughly 200, and dividing it by 40 faculty members, the resultant student–faculty ratio is a comfortable 5 to 1. (Today's is about 10 to 1.) This left plenty of time for off-campus employment, should the opportunity present itself. Depression-era faculty, unlike today's, were largely exempt from the occupational imperatives of grant-writing and continuous publication. Unlike some universities, Columbia did not cut the salaries of its regular faculty, thus rendering them relatively immune to the economic dislocations affecting so many fellow Americans.[34]

No wonder Butler complained that engineering was overstaffed and underworked. "Am I right in thinking we have a much larger faculty than we had twenty years ago (1912)," Butler wrote to Barker in 1932, "when there were registered nearly three times as many students as we now have?" When the president proposed doubling the teaching load of the senior engineering faculty and cutting back on junior staffers, Barker acknowledged that the school's teaching load of eleven hours per week was "light compared to other engineering schools, but not as compared with Columbia schools generally." Nothing came of Butler's proposal and Columbia's aging engineers soldiered on.[35]

To be fair, much of the documentation for this rendering of the interwar engineering faculty is based on the surviving papers of the two put-upon engineering school deans, and the papers of an increasingly beset President Butler. As administrators, it was their job to identify problems, even those for which they had no easy solutions. The profile of the interwar engineering faculty may unfairly highlight instances of departmental dysfunction while failing to do justice to a myriad unrecorded individual instances of academic high-caliber performance. This said, the years of the Great Depression were tough on the nation, on Columbia, and on Columbia engineering.

THE INTERWAR ENGINEERING STUDENT

The first thing to note about Columbia's engineering students of the interwar period was that there were so few of them. Entering undergraduate classes rarely topped fifty members, most already known to each other from two or three pre-engineering years in the College. The few undergraduate transfers were expected to bond with their classmates during the four-week orientation program mounted every August at Camp Columbia in northwestern Connecticut. After modifications were made to the 3/3 plan in 1923, about a quarter of the students enrolled in the 2/2 BS program, a half in the five-year AB/BS program, and a quarter in the six-year professional degree program. Some engineering students came from wealthy families and could look upon a six-year degree program with equanimity and even relief at being off the job market in parlous times. But most Columbia engineering undergraduates came from modest circumstances with little time to spare and securing a paying job their first order of business.[36]

That engineering before World War II appealed to serious students with professional ambitions but limited time likely accounts for the modest role Columbia's engineering students played in extracurricular life. Many held campus jobs or worked part-time. Some participated in intercollegiate athletics, campus journalism, debating, and theater, but not in proportion to their numbers. If the chess club and fencing attracted more than their share, the various political organizations that attracted students of the College and Barnard found few recruits among Columbia's engineers.[37]

Graduate enrollments declined when corporations that had underwritten graduate training in the 1920s no longer did so. Their absence was only partially offset between 1924 and 1932 and after 1938 by the annual enrollment of between twenty and thirty career naval officers pursuing graduate degrees in mechanical and electrical engineering. Some years the navy postgraduate program accounted for upwards of a quarter of the school's graduate enrollments. The program's most famous graduate, then Commander

FIGURE 4.2 Naval engineering officers descending steps of Low Library following commencement, 1917.

Source: Columbian Yearbook, Columbia University Archives.

Hyman Rickover, USN (1930), subsequently admiral and "the father of the nuclear navy," also numbers among the school's most appreciative:

> In 1929 I attended the Columbia School of Engineering for postgraduate study in electrical engineering. Columbia was the first institution that encouraged me to think rather than memorize. My teachers were notable in that many had gained practical engineering experience outside the university and were able to share their experience with their students. I am grateful, among others, to Professors Morecroft, Hehre, and Arendt. Much of what I have subsequently learned and accomplished in engineering is based on the solid foundation of principles I learned from them.[38]

The ethnicity of interwar engineering students mirrored that of the College and Barnard, especially after 1914 when the three undergraduate schools shared a common admissions process. Prior to the 1890s, most engineering students came from New York City's culturally dominant Anglo-Saxon Protestant families, with Episcopalians and Presbyterians in the majority. Jewish students, a small presence into the 1890s, became more numerous after 1905, and despite various policies in place throughout the interwar period to limit the proportion of Jewish applicants admitted to Columbia's three undergraduate schools, they likely accounted for 20 percent of any entering class and a quarter of those who graduated. Anecdotal evidence suggests that between a quarter and a third of undergraduate engineering students were Jewish. Interviews with Jewish graduates of the period produced no instances of overt anti-Semitism. Meanwhile, the proportion of Catholics, drawn mostly from New York's large Irish and Italian communities, quietly rose to the point that by the eve of World War II, Protestant engineering students were likely in a minority. Foreign students, while proportionally more numerous in the engineering school than in the College, made up only about 5 percent of entering classes, and were in most cases Europeans.[39]

The engineering curriculum was, if not more academically demanding, more time-consuming than the College's. Except in the early years of the Butler presidency when charges of widespread cheating were made against engineering students, the impression inferred from the student press is that engineers were a hardworking lot. Even the allegations of cheating speak to the pressure Columbia's early engineering students had placed on them, by family or themselves, to secure professional employment.[40]

The careers these students typically pursued are a further indication of professional caution and risk aversion. Most went to work for large corporations, initially in mining and construction, later for power companies, utilities, and manufacturing. Few struck out on their own. An entrepreneurial ethos was little in evidence. Success came with getting to the top, or at least well up into the technical side of a large corporation such as AT&T, U.S. Steel, or General Electric, or becoming a partner at engineering firms like Parsons Brinckerhoff or J.C. White Construction. A few pre-WWII engineering graduates went on to academic careers but proportionally fewer than did Columbia College contemporaries. A small number took up or

FIGURE 4.3 School of Mines Building, underwritten by mining investor Adolf Lewisohn (1849–1938) and opened in 1907. With opening of Mudd in 1962, renamed Lewisohn and made home of the School of General Studies. Photograph by Margaret Bourke-White, 1931.

Source: Columbia University Archives.

later shifted into careers far removed from engineering, but most promptly, permanently, and cheerfully expected to put their training as engineers to direct use as part of the American professional and corporate workforce.

In November 1939 the Columbia engineering school celebrated its seventy-fifth birthday. While faculty, students, and alumni attending the festivities at the Waldorf Hotel could not have known, for those of us who came later the date has more than a temporal waypoint. It marked the close of one era—the old days—and the opening of another—the modern era— for the engineering school, for Columbia, for American higher education, for the nation, and for the world.[41]

> How poor and miserable we were
> How seldom together
> And yet after so long me thinks
> In those days everything was better.
> —Randall Jarrell, "In Those Days"[42]

WAR, BLESSED WAR

War, blessed war, had come to my generation, and nothing ever would be the same.

—ALFRED KAZIN, *STARTING OUT IN THE THIRTIES* (1977), 56

Unlike the university's response to America's entry into the Great War, when President Butler resisted the interventionist clamor, Columbia's mobilization for World War II began two years before the Japanese attack on Pearl Harbor. And unlike 1914–1917, when campus sentiment was divided between supporting the Allies and staying neutral, following the German invasion of Poland in August 1939, Columbians were in the vanguard of the interventionist movement. Jewish émigrés on the faculty—the sociologist Paul Lazarsfeld, the art historian Paul Kristeller, and the visiting physicists Enrico Fermi, whose Italian wife was Jewish, and the Hungarian Leo Szilard—had personal experience with the devastating impact of the Nazification of German and Italian universities. Others, like the Brooklyn born I. I. Rabi, who had studied physics in Weimar Germany, were similarly persuaded of the reality of the threat posed by Hitler. Unlike Yale, where a contingent of America Firsters opposed aid to the Allies right up to Pearl Harbor, or the University of Chicago, where Chancellor Robert M. Hutchins espoused the isolationism widely held among Midwesterners, Columbia and her septuagenarian president, his brief infatuation with Benito Mussolini behind him, were eager to join the fray.[43]

For some Columbia's scientists, the visitors Enrico Fermi and Leo Szilard, but also the young physicist John R. Dunning and the chemist Harold Urey, mobilization meant joining the larger multi university-based effort to counter what Albert Einstein warned President Roosevelt in August 1939 was Hitler's secret effort to develop a bomb that would utilize the explosive power of a nuclear reaction. For I. I. Rabi, the coming war meant signing on with the MIT-based effort in 1940 to increase the range of enemy detection in the air and on and beneath the sea by means of radar and sonar. For faculty in the social sciences and humanities, it meant offering their knowledge of Germany, Italy, and Japan, as well as America's uncertain ally, the Soviet Union. Others volunteered their language skills and topographical knowledge of such exotic places as the North African coast or South Pacific islands now designated future battle grounds. Both military intelligence agencies and the Office of Strategic Services, headed by Columbia law school alumnus William J. Donovan, found plenty of recruits on Morningside Heights.[44]

Columbia's engineering faculty commenced their contribution to the war effort in 1940 when Dean Barker took up duties as assistant secretary of the Navy. Assistant professor of civil engineering Raymond Mindlin left Morningside soon thereafter for a secret research assignment in Washington. Later, professor of chemical engineering Eric Jette was among the scientists and engineers brought together in 1943 to staff the bomb-making facility at Los Alamos, New Mexico. Several other faculty took on wartime assignments even as they kept teaching.[45]

The first of three substantive curricular responses to the war by the engineering school was to install an accelerated program by which students could complete their studies and secure their diplomas before entering military service. This meant teaching a twelve-month program, abbreviating Christmas break, cutting some instructional corners, and having engineering faculty teach the pre-engineering courses that had been the responsibility of College science faculty. This allowed completion of the traditional four-year BS degree program in two and a half years. But even this did not alter the fact that the School was rapidly running out of students. The imposition of the draft in 1940 for twenty-two-year-old men had an adverse effect on all of higher education but particularly on graduate enrollments at male-only schools. The revised wartime draft enacted in 1942, which lowered the draft age to eighteen, put more downward pressure on enrollments, which at the engineering school dropped by nearly half from a prewar high of 295 in 1940 to 150 in 1942.[46]

A second war-induced adjustment was agreed upon in August 1942 when a majority of the engineering department chairs voted to have the school admit women, beginning in February 1943. Acting Dean Finch, despite acknowledging that enrollments "were extremely low" and that the very continuation of the school was in question, opposed the resolution. Besides a more general identification with the unbroken and only twice questioned eighty-year prohibition against admitting women, he voiced two specific objections. "It is easy to grant this so-called right to women," he allowed, "but it will probably be impossible to withdraw it should we desire to do so at a later date." He also thought that few women would apply and cited the low numbers of women at prewar co-ed engineering schools.[47]

Finch was right on both counts. The 1942 statute allowing admission of women was not limited to the war years, thus ending the engineering school's dubious distinction of being Columbia's only professional school to exclude women. And few women applied, either during the war or in the two decades that followed. But the first to do so, Gloria Brooks (later Gloria Reinish), a seventeen-year-old transfer applicant from Cooper Union, promptly did so. Upon admission in February 1943, with full scholarship and advanced standing in electrical engineering, she soon found herself in classes filled with sailors in uniform. Aside from a single instance of hazing, she was warmly accepted by her classmates. Barnard College proved to be a source of "some little indignities," not allowing her to attend Barnard dances or use its tennis courts. As for Acting Dean Finch, who told her of his opposition to admitting women, "he was a perfect gentleman."

FIGURE 4.4 Gloria Reinish (née Brooks), in 1943, the first woman to be admitted to the Columbia engineering school. Degrees include BS 1945, MS 1948, PhD in chemical engineering 1974). Mother and grandmother of SEAS graduates; professor of biomedical engineering Fairleigh Dickinson University (1973–).

Source: Columbia University Archives.

Reinish graduated with honors in 1945 (but not Tau Beta Pi, then closed to women) with a degree in electrical engineering. While working at Sperry Gyroscope, she returned to Columbia in 1948 for an MS degree, when she encountered in the persons of John Ragazzini, Morton Arens, and Ralph Schwarz "really, really great teachers." As for Mario Salvadori: "Wonderful." After raising three children, she returned to Columbia and completed a PhD in biomedical engineering in 1974, under the direction of Edward Leonard. Leonard excepted, she found the teaching this time "not as good" as earlier. By then she was a tenured member of the faculty of Farleigh Dickinson University, a position she continued to hold when interviewed in 2012. In the interim, a son, a daughter, and a granddaughter of Professor Reinish number among SEAS's graduates.[48]

Inclusion in the Navy's V-12 program had a bigger impact on the school than either the accelerated program or the admission of women. Of all the military services, the wartime Navy was the most determined to staff its commissioned ranks with college graduates. Advanced technical schooling was viewed as especially crucial for those on shipboard assignment as engineering and telecommunications officers. Accordingly, early in the war the Navy set about identifying universities ready to meet its educational needs. Columbia's engineering school, with its long history of providing graduate training to career naval engineers and its already-in-place accelerated undergraduate program, was a natural candidate. Lobbying by Special Assistant to the Under Secretary of the Navy Dean Barker ensured Columbia's inclusion among 131 other colleges and universities in the V-12 program, which was launched in the summer of 1943.[49]

The first five hundred sailors arrived on the Columbia campus in July 1943. They consisted of enlistees right out of high school, others who had been inducted out of college, and the remainder those already in uniform with some college training. The latter were given advanced standing and were expected to complete a degree, and with it an ensign's commission, in nine to fifteen months. Fifty or so of those coming to Columbia with prior college math and science training were admitted directly to the engineering school. Thirty-three sailors joined twelve civilians in the February 1944 graduating class, while two hundred fifty V-12ers were enrolled in the engineering school, and another two hundred fifty enrolled in the college's pre-engineering program. Many of these first graduates were assigned shipboard duties in the closing months of the European and Pacific campaigns. Four died in action.[50]

Engineering enrollments reached a wartime high of six hundred students in the spring term 1945. Most of the V-12ers were still on campus when the war ended and were allowed to complete their degrees before accepting a commission in the peacetime Navy or returning to civilian life.

Among the nearly one thousand graduates of the V-12 program in engineering, and the even larger numbers to have graduated from Columbia College and the medical school, are included hundreds of Columbia's most devoted alumni.[51]

As the war wound down in the summer of 1945, Acting Dean Finch voiced concern about what he considered the inevitable return to the prewar status quo. The loss of ensured enrollments of between five and six hundred well-disciplined young men, whose tuition (and then some) had been promptly and fully paid by the American taxpayer, was not a happy prospect. Of equal concern was whether younger faculty away in military service would be returning. Whether or not a consolation for the acting dean, these looming administrative problems were not his to confront. They would the job of Dean Joseph Barker, finishing up at the Navy Department and expected back on campus for the fall term. And who better than a war-tested administrator familiar with Washington to meet the challenge of adapting Columbia's engineering school to the demands of the dawning "American Century"?

5

MISSING THE BOAT 1945–1964

You can count on our not forgetting about Engineering.

—ACTING PRESIDENT FRANK FACKENTHAL

TO DEAN JAMES KIP FINCH, FEBRUARY 7, 1945[1]

We are forced to sit back and make no effort to secure government or other
research funds—we have no place for them. . . . We have missed the boat.

—DEAN JAMES KIP FINCH TO ACTING PRESIDENT

FRANK D. FACKENTHAL, JANUARY 28, 1948[2]

AMERICANS IN THE two decades following World War II came to hold institutions of higher education in singularly high regard. This was particularly true of the nation's research universities, both public and private, and even more of the half dozen leading universities that enthusiastically contributed their talents to the nation's ongoing struggle with the Soviet Union and international communism. Expertise came from the humanities and social sciences, which provided a counternarrative to Marxist thought, to burnishing the image of American "exceptionalism," to advancing a historical case for liberal anti-communism, and to the development of foreign area programs and foreign language studies. Still more conspicuous were contributions from science and engineering departments to the cause of national defense, development of advanced weaponry, enemy detection, and, when it too became a venue for competing with the Russians, space exploration. Universities eagerly enlisted in the Cold War just as they had in the successfully concluded "crusade" against fascism.[3]

More than patriotism was involved. Economic and reputational factors played a part. To a large extent, the ability of universities to meet the government need for expert assistance determined their place in the competitive

reordering of postwar universities. Some universities and engineering programs, by the magnitude of their contributions, went up the reputational
ladder, others held their place, while others lost standing. When Dean
Finch declared in 1948, "We have missed the boat," he referred specifically
to his beloved engineering school. If any consolation, much of Columbia
remained dockside, too.[4]

THE LEADERSHIP VACUUM AND
THE SPACE IMPERATIVE

In March 1946, six months back at Columbia after five years in Washington,
Joseph Barker resigned his position as dean of engineering to become president of the Research Corporation. Finch again stepped into the breach, this
time as the school's seventh dean. At sixty-three, and after thirty-seven years
as a member of the engineering faculty, the last eight as acting dean, he was
viewed by most observers (though not by himself) as an interim appointment. What the school was generally thought to need at this juncture was
not someone who came of professional age back before World War I and
whose recent wartime service consisted of keeping school on the home
front. Needed was a dynamic dean attuned to the new academic opportunities of the postwar era. Needed was a Columbia equivalent of MIT's
Vannevar Bush or Stanford's Frederick Terman, and the diffident Finch was
neither. Barker, freed from the financial exigencies of the Depression and
experienced in the ways of wartime Washington, might have served, but he
had bailed.[5]

New leadership was all the more crucial at the school level because of
the extended leadership vacuum existing at the top of the university. Since
the trustee-mandated retirement of the eighty-three-year-old Butler in
May 1945, the university had been operating under the authority of Acting
President Frank D. Fackenthal. Fackenthal had begun his career at Columbia in 1904 as a secretary to the president, progressing through the ranks

until in 1937 Butler restored the position and made him university provost. As the university's ranking academic officer for eight years and then for two years as acting president, the sixty-four-year-old Fackenthal proceeded cautiously.[6]

Butler's departure negotiations specifically excluded his longtime staffer from consideration as his successor. This suited the trustees, especially trustee and IBM head Thomas Watson, who in 1947, while not on the presidential search committee, personally convinced the unemployed general of the Army Dwight D. Eisenhower to accept the Columbia presidency, effective July 1, 1948.[7]

The leadership vacuum did not end with Ike's arrival. Disappointed in finding Columbia not a country college in the Williams/Amherst mold and unsettled by the faculty's casual disregard of rank, he never fully unpacked. Eisenhower spent most of his five years as president of Columbia back in uniform in Europe as supreme commander of NATO and campaigning as the Republican nominee for the 1952 presidency.[8]

The most pressing problem the engineering school and its dean faced in the immediate postwar era was space. Unlike some peers, Columbia came out of World War II with no new federally funded real estate to show for it. MIT substantially increased its laboratory space by the creation of the Radiation Lab adjacent to the main campus in Cambridge and the Lincoln Lab in the Boston suburb of Weston; the University of Chicago added a federally funded metallurgical laboratory; Berkeley acquired its cyclotron on the government's nickel; Cal Tech and UCLA gained access to the nearby Jet Propulsion Laboratory. All these facilities provided modern research space for returning faculties and students. By contrast, Columbia's postwar engineers were still confined to the cramped and dispersed spaces in use since the 1920s, all showing their age. Metallurgy professor Daniel Beshers recalled the situation in the School of Mines building on his arrival in 1957: "The paint was peeling off the walls in great pieces, perhaps a quarter inch thick and 10 feet or so across."[9]

Lack of space had been a problem raised by every engineering dean back to Chandler. Dean Goetz alluded to it in 1910 when informed that the adjacent site south of the Mines building fronting on Broadway would no longer be "a future addition to the technology group" but would go instead to the new journalism school. In the 1920s Dean Pegram pushed a never-acted-upon plan to relocate the civil and mechanical engineering departments to an off-campus site adjacent to the Harlem Ship Canal. Part of Dean Barker's frustration upon arrival in 1930 was the cramped, dispersed, and aging quarters allotted to the school.[10]

This space crunch a half century after the move to Morningside was university-wide. Back in 1903, when the land surrounding the new campus remained undeveloped, President Butler suggested to trustee and banking magnate J. P. Morgan that the university should buy up surrounding properties for future development. Not necessary, the financial titan assured him: "It will all be here for the buying whenever Columbia needs it." Five years later, with the IRT subway linking downtown Manhattan to the Upper West Side, the once vacant blocks surrounding the campus quickly filled with ten- and twelve-story residential apartments that remain today. Aside from the Medical Center, opened on Washington Heights in 1928, nearly all Columbia's new construction between 1897 and 1945 was contained within the original four-block site and the two city lots directly to the east. No construction had occurred on the campus since completion of Butler Library in 1935.[11]

In 1945, Acting Dean James Kip Finch declared the engineering school's plant totally inadequate if Columbia hoped to compete in the postwar period. Two years later he described the school's situation to Acting President Fackenthal: "We are forced to sit back and make no effort to secure government or other research funds—we have no place for them."[12]

Fackenthal agreed that the situation required action and endorsed Finch's idea for a capital campaign for the engineering school. Both knew, if approved by the trustees, such a campaign would be a university first. But with Butler no longer there to oppose fundraising directed at alumni, the trustees endorsed a capital campaign on behalf of the engineering school. President-elect Eisenhower, although specifically exempted from fundraising in his deal with Watson, concurred.[13]

Having originated the idea of a capital building campaign, Dean Finch might have seemed the obvious choice to lead it. Not so. One of Ike's first presidential decisions upon arrival, urged by his wartime staffer and University Director of Fundraising Paul H. Davis, was to make a change in engineering deans. To this end, an amendment was made to the University Statutes making mandatory the retirement of any dean at age sixty-five. It fell to University Provost Albert C. Jacobs to tell Finch that his next year at Columbia would be the last. "It is my sincere hope," Jacobs wrote on March 25, 1949, after quoting the amendment to the Statutes that made the dean's severance "automatic, clear, complete and final," "that this retirement is not unwelcome and that the period of notice is adequate for your plans."[14]

Finch went less quietly than his mild demeanor suggested. He took personal umbrage at the "new rule" devised to force his retirement. As for his thoughts on a suitable successor, he gave the search committee his unsolicited views:

1. He be "a Columbia man";
2. He have broad ties to industry;
3. He be an engineer—not a pure scientist;
4. He be someone the engineering faculty and alumni wanted—they had proposed General Lucius D. Clay (then finishing up as military governor of the U.S. zone, Germany);
5. He not be Professor of Physics John R. Dunning, even on a temporary basis.[15]

It was to no avail. On July 1, 1950, John R. Dunning was installed as the Engineering School's eighth dean.[16]

"OUR NOBEL LAUREATE"

A Nebraskan who had come east in 1930 to study physics at Columbia with Pegram, John R. Dunning became one of the physics department's prize PhD students. He became an instructor in 1933 and completed his doctorate a year later. In 1935 he was appointed assistant professor; the following year he used a Columbia Traveling Fellowship to discuss his research in neutron physics with Europe's leading nuclear physicists. Back on Morningside in 1937, he built a cyclotron in the basement of Pupin Hall. Working with Columbia physicists I. I. Rabi, Willis Lamb, and Enrico Fermi, all three subsequent Nobel laureates, Dunning was one of the first Americans to test the operational implications of nuclear fission after it had been confirmed experimentally on January 13, 1939. Twelve days later, using the Dunning-constructed cyclotron in Pupin, Fermi and Dunning led a six-member experimental team that conducted the first successful nuclear fission experiment in the United States. All six went on to play important roles in the Manhattan Project and the wartime development of the atomic bomb.[17]

Following the removal of the Manhattan Project from Columbia in 1941 to the University of Chicago, Dunning became director of a gaseous-diffusion lab at Hanford Washington. He was subsequently transferred to Los Alamos, where he made a crucial contribution to the separation of uranium isotopes. He was present for the first test of the atomic bomb in the New Mexico desert on July 16, 1945. For the secret work he did during the war, Dunning received $300,000 from the federal government in lieu of patent royalties. Few Columbians had such an exciting or rewarding war.[18]

Back at the Columbia physics department in 1946, Dunning commenced a loose connection with the engineering school through his appointment as the first Thayer Lindsley Professor of Applied Science. He became the founding director of the Nevis Laboratories, in Irvington, New York, a cooperative endeavor of Columbia, the Atomic Energy Commission, and the Office of Naval Research, where classified research was carried out using what at the time was the world's largest cyclotron. In 1947 Dunning was awarded the Medal of Merit by President Truman. A year later he was elected to the National Academy of Sciences. Unlike several of his Columbia physics colleagues, Dunning never received the Nobel Prize. But in his later years, when masters of ceremonies at engineering alumni functions took to referring to him as "one of the world's outstanding nuclear physicists" or "our Nobel laureate," he did not demur.[17]

FIGURE 5.1 John R. Dunning (1907–1975) (right), physicist, promoter of nuclear energy, and seventh dean of Columbia engineering school (1950–1968) with his predecessor, physicist George B. Pegram.

Source: Columbia University Archives.

Dunning's bull-in-a-china-shop personality marked him in some Columbia minds as too much the self-promoter. He certainly never minimized his role in the making of the atomic bomb or championing the postwar nuclear power industry. It has also been suggested that I. I. Rabi, chairman of the Columbia physics department and someone with Eisenhower's ear, may have urged Dunning's appointment as a way of easing him out of Pupin Hall. But others, including President Eisenhower, saw him as a man of action and charisma, just the sort of war-tested leader to revive Columbia's moribund engineering school and energize its forthcoming capital campaign.[19]

Dunning took little interest in the day-to-day workings of the school, leaving these to his associate dean, Wesley Hennessy, a nonengineer and nonacademic, whom Finch had hired back in 1942. He served as a consultant and publicist for the nuclear power industry, while also on call to the Atomic Energy Commission. Faculty there in the 1950s recall his big cigars, power suits, and feet-on-the-desk manners. His frequent absences from the office were attributed to corporate board meetings and fundraising, but nobody knew for sure. Professor of metallurgy-emeritus, Daniel Beshers, remembers Dunning as "an almost absentee dean. From my lab on the sixth floor of Mines I could look down to his office window on the main floor of Engineering (now Math) and see that he often came in about 5 P.M. and was opening his mail."[20]

Dunning operated during the last two years of the Eisenhower presidency and early in the presidency of Grayson Kirk, who succeeded Ike in 1953, with little effective oversight. William J. McGill, Columbia professor of psychology in the 1950s, later characterized Dunning and Business School Dean Courtney Brown as "the barons who achieved such success for themselves and such a bad reputation for Columbia for weak administration." On Dunning specifically:

> I thought he was a terrible dean of engineering, because he was a wheeler and dealer, a man without any engineering outlook, and someone who was willing to feather his own projects. As a result, a number of prominent departments in the School of Engineering went downhill during his stewardship.[21]

Other contemporary assessments were more positive. President Eisenhower and, for a time, President Kirk, spoke highly of Dunning's fundraising prowess. (Dunning later credited himself with raising $50 million, an exaggeration that included gifts that came despite his being dean.) Then again, he met the expectations of engineering alumni who wanted a dean comfortable among downtown movers and shakers and on a first-name basis with generals and admirals, whatever his shortcomings at keeping school. That he made little impact on students and had no contact with junior faculty only confirmed the import of their dean.[22]

TRIGA AND ITS DISCONTENTS

In the early 1950s, the destructive character of nuclear energy made manifest at Hiroshima and Nagasaki, the possibilities of its peacetime uses became apparent with the first operational nuclear reactors. The potential of nuclear power for generating electricity made it attractive to public and private investment, and in turn for research and instruction. At Columbia, as elsewhere, faculty divided between those identified with the Manhattan Project during the war and the postwar research related to the production of nuclear weaponry, and those who took up the challenge of producing nuclear energy for peacetime purposes. In either case, newly available funding from the Atomic Energy Commission (AEC) and Department of Defense became attractive to universities coming to rely on federal research funding to underwrite non-income-producing operations.[23]

Columbia numbered prominently among these "federal research universities" and included on its postwar faculty several nuclear physicists and nuclear engineers; the most prominent was Dunning. Other engineering faculty with a professional investment in nuclear research included the applied physicist William Havens; chemical engineer Carlos Bonilla; mechanical engineers Herbert Goldstein, Leon Lidofsky, and Edward Melkonian; and the applied physicist Robert A. Gross. But more than talented professors was required to build a credible university nuclear engineering program. Needed also, or so the conventional wisdom dictated, was a university's own nuclear reactor. Nuclear physicists and engineers by the mid-1950s at some fifty universities had access to nuclear reactors: Why not Columbia?[24]

In point of fact, Columbians did have access to a nuclear reactor in Brookhaven, built by a consortium of northeastern universities on the eastern end of Long Island, seventy miles from Manhattan. Columbia physicist I. I. Rabi had been a prime mover in securing the federal funding to build it. The Brookhaven reactor became operational in 1950. In addition, federal funds secured by the Columbia physics department in the 1950s underwrote the cyclotron housed in the department's Nevis Labs, located in Westchester County, twenty-five miles north of Manhattan.[25]

Neither facility met the stated needs of Dunning, who in 1960 got university approval to apply for $250,000 from the National Science Foundation to purchase a new version of a TRIGA Mark II reactor for education and research. In May of 1963 a second application to the AEC for construction of the facility to house the reactor was submitted and approved. Construction began in 1964 and the reactor was installed in the Engineering Terrace basement three years later. In January 1968, on the basis of assurances that the automatic shutdown features built into TRIGA eliminated the possibility of the reactor overheating, supposedly rendering it safe even if located in one of the nation's most densely populated neighborhoods, the university applied to AEC for an operating license.[26]

This followed a protracted series of safety tests by city, state, and federal agencies to provide assurances of the reactor's safety. Meanwhile, neighborhood groups and local politicians, already exercised by Columbia's real estate policies, which seemed designed to clear its surrounding neighborhoods of what University Provost Jacques Barzun characterized as "its dangerous elements," declared their opposition to the presence of TRIGA and to its activation. The anti-TRIGA campaign garnered support from student organizations and some faculty, mostly in the humanities, but also a few in the pure sciences, who framed their opposition in terms of residential proximity to the reactor.[27]

Those favoring activation included the university trustees, President Kirk, Dean Dunning, and all but two members of the faculty of SEAS, plus the engineering student newspaper, *Pulse*. A few nonengineering faculty viewed it as an academic-freedom issue allowing faculty to engage in research of their choosing, and supported activation. Yet even with all the safety certifications in hand in early 1968, the tumultuous campus events of that spring and the two following springs led university administrators to defer activation until a hoped for quieter time, leaving the issue still unresolved when William J. McGill became president in July 1970.[28]

By then, Dunning was no longer dean, having been all but fired in 1964. Over a weekend in May, the trustees voted by mail ballot to strip Dunning of operational responsibilities for the Engineering School, consigning him to an out-of-the-way office in the basement of Engineering Terrace and limiting his duties to alumni relations. On July 1, 1964, Dunning's longtime associate dean, Wes Hennessy, became "executive dean," with operational responsibility for the school.[29]

The precise reasons for this abrupt action remain a mystery. Faculty at the time have surmised that it involved Dunning's efforts to secure tenured appointments for research staffers at one of his favored projects, the Electronics Research Lab located on 125th Street, over the opposition of the electrical engineering department and without the knowledge of the president. Whatever the reason, Dunning was now dean-without-portfolio and remained so for five years until reaching early retirement age in 1969. Hennessey then became the engineering school's ninth dean. Dunning retired to Florida, where he died in 1975.[30]

THE KRUMB MILLIONS: LIFELINE OR SEA ANCHOR?

Another two-edged legacy of the Dunning deanship was the Krumb benefaction. Upon graduating from the School of Mines in 1898 with a degree in mining engineering, young Henry Krumb went west to make his fortune. Moving around Arizona, New Mexico, and Colorado in the employ of New Yorker Meyer Guggenheim, he applied advanced techniques to the mining and assaying of copper and silver, eventually becoming a principal in the American Smelting and Refining Corporation, of which Columbia benefactor Adolph Lewisohn was the major stockholder, and later head of his own equally profitable Pelmont Mining Corporation. He returned to New York in the late 1930s and became active in the Engineering School Alumni Association, giving $35,000 in 1940 to establish "The Henry Krumb Scholarship Fund," in recognition of the school's having provided him a scholarship of $200 without which he would have had to leave school in his senior year. In 1941 he was elected one of the board's six alumni trustees and served the allotted six years.[31]

FIGURE 5.2 Henry Krumb (1875–1958) (EM 1898), Columbia-trained mining engineer and investor; gave engineering school $16 million to upgrade and perpetuate its mining program. At announcement of the gift in 1948, the largest ever given to Columbia.

Source: Columbia University Archives.

In 1948, on the occasion of the fiftieth reunion of the Class of 1898, Krumb revealed his intention to leave the bulk of his estate to the engineering school. Half—some $6.6 million—was to go to the engineering school at his death, the other half upon the death of his wife, La Von. Estimates placed the combined gifts in the range of $13 million.[32]

However generous, Krumb proved to be an exacting donor. Rather than have his money go to Columbia University or even its engineering school for general purposes, he stipulated that it to go to making those parts of the engineering school traditionally linked to the mining industry, the Department of Mining, Metallurgy and Mineral Engineering, into "the world's best school of mines." The fact that mining no longer occupied the prominent place it had in the school's operations, and that the faculty in metallurgy and mineral engineering had long since branched out into research areas removed from mining as Krumb had experienced it earlier in the century, was not the would-be donor's problem. On an inspection tour of the School of Mines building in the 1950s, Krumb asked, "Where are the assaying tables?" unaware that the use of such mine-site equipment had been dropped from the curriculum thirty years earlier.[33]

The actual negotiations with Krumb were assigned to Professor of Mining Menelaos Hassialis. Dunning tried to interject himself into the discussions, but backed off when his bluster about the "Krumb Towers soaring over the Columbia campus" elicited a public rebuke from the would-be donor. President Kirk, who by his own account "worked very hard to bring [the Krumb gift] to us," tried to persuade Krumb to leave the specific use of his money to university officials. No such luck. "Henry was a very strong-willed person, and he was deeply interested in the school," Kirk later recalled. "I had to withdraw my own proposals." One last presidential attempt to stir some of the money away from the Department of Mining, Metallurgical and Mineral Engineering provoked Krumb to warn him: "If you insist on doing it this way, I'm going to change my will." Kirk: "At this point I backed down."[34]

The terms of the gift show the lengths to which Krumb went to avoid his monies being diverted to other purposes than those he stipulated.

FOURTEENTH: I give and bequeath the sum of Five Hundred Thousand ($500,000) Dollars to THE TRUSTEES OF COLUMBIA UNIVERSITY IN THE CITY OF NEW YORK as an endowment to establish and maintain the Krumb Chair of Mining, the income of said fund to be applied toward the payment of the salary of a Professor of Mining and for no other purpose.

In acknowledging that $500,000 was well over the going rate for an endowed chair, Krumb made clear his restorative intentions:

> As I wish to again make the Mining Department the outstanding department of the School of Engineering, which it once was, I consider it important that it should be headed by a nationally known Mining Engineer and administrator and, in order to secure such a man, it will be necessary to pay him substantially more than Columbia has ever paid its Professor of Mining.[35]

Krumb did not rely on admonitory pronouncements alone to see his will be done. Elsewhere he exacted agreements from the trustees that they would make annual reports on the two principal funds—"The Henry Krumb Mining and Metallurgical Scholarship Fund" and "The Henry Krumb Mining Engineering Fund"—as to the uses the income had been expended. The first of these reports was to be shared with the trustees of Memorial Center for Cancer and Allied Diseases [now Sloane-Kettering], the second with the trustees of the Lenox Hill Hospital. The will further designated these hospitals as the default recipients of half of whatever Krumb monies remained should Columbia ever be found to have used these funds for purposes unspecified by the donor.[36]

If several terms of the gift were so prescriptive as to question why the trustees acceded to them, one is notable for its magnanimity. This involved simultaneous negotiations with the trustees of the estate of Seeley W. Mudd, a successful mining engineer who died in 1926, but not before taking a philanthropic interest in engineering education, especially in Southern California, where he became an important benefactor of Pomona College. Mudd's executors agreed to contribute $1 million to the capital campaign toward the cost of the planned engineering school building, but only if the building, which was expected to cost $6 million, was named "The Seeley Wintersmith Mudd Building." Informed of this development, Krumb agreed that half of his residuary estate ($3,000,000) "be applied toward the cost of the proposed New Engineering building," waiving any naming claims.

Even here, Krumb could not resist: "If the new building does not have materially more usable space than is at present in use, some of the space in the old Engineering Building now used should be retained in addition to the new building."[37]

The total Krumb benefaction of $16,000,000 was at the time the largest gift ever received by Columbia University from a single donor. In addition to paying for half the cost of Mudd, which opened in 1962, it covered most of the cost of the smaller Engineering Terrace Building that followed two years later. The gift also underwrote a total of twenty-one faculty positions, evenly distributed among the mining, metallurgy and mineral engineering divisions of the newly named "Henry Krumb School of Mines, Department of Mining, Metallurgical and Mineral Engineering." For many years one of the smaller engineering departments, the least funded and most under enrolled, it now became the engineering school's largest department, its richest, best housed, with four floors devoted to it, but still its least enrolled. The Krumb chair made its holder the second highest paid professor in the university.[38]

For such good fortune to be visited upon a single department without some grousing from less blessed quarters defies academic reality. Faculty in the science departments, dependent upon incessant grant-writing to secure their keep, regarded such largesse unmerited. Engineering faculty complained of the unfairness of this gift, while less well-heeled departments, to whom fell the bulk of the school's teaching, went without. One faculty member has since allowed the possibility that the long-term effect of the Krumb legacy has been less to provide the struggling engineering school with "a life line than a sea anchor," holding the school back from moving in more productive directions. One of the endowment's most protective beneficiaries, Professor of Mines Colin Harris, allowed toward the end of his career that the Krumb monies may have had a distortive effect on the school overall. Questions about the utilization of the Krumb millions continued to bedevil successive engineering deans into the twenty-first century.[39]

Yet the beneficial effect of the Krumb gift on the school overall should not be discounted. Without it the 1950s capital building campaign would not have reached its goal and the chronic space constraints that had plagued the school since the 1920s would have persisted. Several of the faculty in Krumb-endowed positions, including National Academy of Engineers members Herbert H. Kellogg and Nathaniel Arbiter, and others recruited with funds from the Krumb endowment, among them Daniel Beshers, Paul Duby, Arthur S. Nowick, Ponisseril Somasundaran, and Nicholas J. Themelis, the last three also members of NAE, have figured among the school's most distinguished researchers and teachers.[40]

Worth mentioning are three other points regarding the controversy surrounding the Krumb gift:

1. The idea that President Kirk or the trustees, or any president or Columbia board in the midst of a struggling capital campaign would have seriously considered turning down the largest potential gift in the university's history because of donor specifications of the sort Krumb made is hardly credible;

2. The fact that the engineering school's restricted endowments in 1972 totaled $22.5 million, thanks mostly to the Krumb millions, made it second only to the medical school among Columbia's schools in the size of restricted endowments, thus providing ample insurance in the early 1970s against the possibility of closing the school for financial reasons;[41]

3. The school's viability five decades after receipt of the Krumb millions indicates, however temporarily distortive their impact, that they did not prove permanently disabling. Thus ends the lesson on looking this particular gift horse in the mouth.

RECONSTITUTING A POSTWAR FACULTY

In addition to the space problem, the engineering school in the immediate postwar years confronted a depleted faculty. Faculty shortages were widespread throughout Columbia, but the engineering school's were especially severe because it had gone through the Depression years with a virtual hiring freeze and had held back on hiring due to stagnant enrollments. The result was a top-heavy age distribution that had a quarter of the entire faculty (ten of forty) leaving between the start of the war and its end, seven by retirement, three by death. The faculty in 1945 was 25 percent smaller than six years earlier, down from forty to thirty-two.[42]

The impact of these losses differed by department. Of the seven mechanical engineers on the faculty in 1940, only Theodore Baumeister and Carl F. Kayan remained in 1945. The chemical engineering department lost four of its seven 1940 members by 1948; industrial engineering, one of three; electrical engineering, three of eight; mining, metallurgy and mineral engineering, two of six. Only civil engineering emerged from the war in reasonably good shape to meet with the postwar challenges. Among the in-place civil engineers were Jewell M. Garrelts, at midcareer; the young Raymond Mindlin, who joined the department in 1940 and spent the war years engaged in secret research; and Nelson S. Fisk, plus veterans William J. Krefeld and Donald Burmister.[43]

Electrical engineering benefited from two junior hires it made on the eve of the war, John Ragazzini, assistant professor, and Ralph J. Schwarz, an instructor, both of whom would play important parts in the department's future. For departments like civil and electrical engineering, selective hiring became the immediate order of the postwar years; for the others it was more a case of starting over.[44]

And start over they did. Between 1945 and 1952 the engineering school averaged three new appointments a year and grew by 63 percent. By 1955, the faculty had grown to sixty-eight members, only twelve of whom had been there a decade earlier. It was not only half again as large as it had been in 1939, it was younger by seventeen years on average. In 1939 every SEAS faculty member was over forty; in 1955 half the faculty was under forty.[45]

In other ways the 1955 faculty resembled their interwar counterparts. Almost half were native New Yorkers, with another quarter coming from the northeast. The proportion of foreign-born faculty was higher than before the war, but still exclusively Europeans and Canadians. The first non-European, the Russian Iranian electrical engineer Lotfi Zadeh, joined the faculty in 1950.[46]

The 1955 faculty also shared with their earlier counterparts a heavy concentration of members trained at Columbia. Over half of the 1955 SEAS faculty received their graduate training at Columbia; the next largest contingent was from MIT (9 percent). Nearly all were hired as assistant professors, often prior to acquiring a PhD. Most went on to secure tenure and those who did mostly stayed on at Columbia. Securing tenure became less certain for faculty hired in the early 1950s but not by much. One member of the mechanical engineering department tenured in the early 1960s later told a junior colleague in the 1980s that the scholarly expectations for tenure back then could be met with "seven published papers," approximately what the school might expect today of a starting assistant professor.[47]

A GOLDEN AGE?

Given this demographic profile, just how good was this post-WWII generation of engineering faculty? Or more precisely, how valued were the contributions they made to their engineering fields? While any effort to assess quality is necessarily subjective, the faculty of 1955–1956, midway through the postwar era, provides a useful sample. Most of its members enjoyed professional lives extending into the 1980s and were thus of an era to be eligible for election to the National Academy of Engineering, a serviceable metric by which reputational standing can be gauged.[48]

Of the sixty-eight men who made up the 1955–1956 engineering faculty, thirteen were later elected to the National Academy of Engineering. All but one of the school's then seven departments—chemical (two), civil (six), electrical (two), mechanical (one), and mines (two)—had at least one future NAE member. When nearly one of every four members of the faculty would later be so recognized as leaders in their fields, claims to the school having experienced a postwar "golden age" must be taken seriously.[49]

Ferdinand Freudenstein joined the mechanical engineering department as an assistant professor in 1954. An Austrian émigré of Jewish parentage, he migrated first to England and then to New York as a teenager in the late 1930s. After graduating from Harvard in 1948, he came to Columbia for graduate studies. In his PhD dissertation, Freudenstein developed what was to become the "Freudenstein Equation," which uses a simple algebraic method to determine the position of an output lever in a linkage mechanism. He remained at Columbia throughout his career, where he made a revolutionary contribution by applying digital computation to the kinematic synthesis of mechanisms. His own research and that of his five hundred PhD students later earned him the sobriquet, "The Father of Kinematics," which he defined as "the mathematical investigation of the motions that take place in mechanisms and machines and the investigation of the means for creating these motions." He continued as Higgins Professor of Mechanical Engineering until his death in 2006.[50]

Columbia's electrical engineering department in the mid-1950s was home to two future National Academy members. Lotfi Zadeh, a Russian-born Iranian, came to Columbia in 1946 for graduate studies under the direction of John Ragazzini. Upon completion of his PhD in 1949, he became an assistant professor. With Ragazzini, Zadeh is credited with having pioneered in the development of the z-transform in signal processing, now standard in digital signal processing, digital control, and discrete-time systems used in industry and research. He later became known as one of the first exponents of what he called "fuzzy logic," a non-Euclidean approach to calculating probabilities. After a decade teaching at Columbia, he left to become the first chairman of the computer science department at the University of California, Berkeley. He returned to Columbia in 2011 to receive the engineering school's Egleston Medal.[51]

The other future NAE member of the 1955 electrical engineering department, Cyril M. Harris, became an assistant professor in 1952. A California native with a PhD from MIT, Harris, during his thirty-two-year career (he retired in 1984) as a member of the department and the School of Architecture, became a world leader in acoustical engineering. He served as acoustical consultant for the construction of the John F. Kennedy Center for the Performing Arts in Washington, D.C., and the Lincoln Center Metropolitan Opera House. He died in 2011.[52]

The two future NAE members in chemical engineering came to Columbia by different routes. Thomas B. Drew, a product of MIT, came to Columbia in 1940 as department chairman. He became widely known for introducing the first systematic use of heat, mass, and momentum fundamentals in industrial applications. His 1983 election to NAE cited him for "pioneering contributions to heat-mass transfer and nuclear engineering technology, educational leadership, and outstanding service to government and his profession." He returned to MIT in 1966. In 2008 he was posthumously named by the American Institute of Chemical Engineering one of the fifty most prominent chemical engineers of the discipline's "Foundation Age."[53]

Elmer L. Gaden came to Columbia as a V-12 student in 1944 and stayed on as a graduate student and instructor until being appointed assistant professor of chemical engineering in 1954. His early work involved determining the amount of oxygen necessary to fuel the fermentation process; he later adapted his findings to facilitate the large-scale manufacturing of penicillin. In 1971 *Engineering and Chemical News* called him "The father of biochemical engineering." Three years later he left Columbia for the University of Vermont. From there he moved to the University of Virginia, from which he retired in 1994. In 2009 he became the fifth recipient of the Fritz J. and Dolores H. Russ Prize, the world's top honor in bioengineering. He died in 2012.[54]

Two of Columbia's future NAE members in 1955 were metallurgists in the department of mines, metallurgy and mineral engineering. Herbert H. Kellogg joined the engineering school as an assistant professor in 1946, after earning both his BS and MS at Columbia. In 1956 he became the second Stanley Thompson Professor of Chemical Metallurgy. He was a leading expert in reducing waste and runoffs at mining sites and increasing the efficiency of mineral processing. Following retirement from Columbia in 1990, he continued to write about the responsibility engineers have for ensuring a sustainable environment. Kellogg's colleague, Nathaniel Arbiter, another Columbia product, joined the engineering faculty in 1951 after several years in the mining industry in Arizona. He remained an active consultant until his retirement from Columbia in 1977, the year of his election to NAE, when he returned full-time to the mining industry. He died in 2008.[55]

Of the school's thirteen future members of the National Academy of Engineers on the 1955 faculty, six were in what was then a twelve-person civil engineering department. First in seniority and prominence, a member of both the National Academy of Engineering (1966) and the National Academy of Sciences (1973), was Raymond D. Mindlin, a native New Yorker and Columbia graduate (AB 1930, BS 1931, EM 1932, PhD 1936) who became an assistant professor of civil engineering in 1939, by which time he had already achieved a measure of fame for using his dissertation to solve what became known as the "Mindlin Problem," relating to classical three-dimensional elasticity in applied physics. He was an early recipient of the von Karman Medal of the American Society of Civil Engineers in 1961 and the Stephen Timoshenko Medal in 1964. Over the course of his career he authored and coauthored 129 papers and sponsored dozens of PhDs. Despite a standing offer from Harvard, Mindlin stayed on at Columbia until retiring in 1976. He died in 1987.[56]

Mindlin's colleague, Mario Salvadori, was born in Rome, Italy, and came to New York in 1938, having completed his PhD at the University of Rome in 1933. He became an instructor in civil engineering at Columbia in 1943. For fifty years he taught in both the engineering school and the School of Architecture until he retired in 1993. He served as consultant responsible for the structural integrity of several famous building complexes in New York City, including the CBS Building, and others throughout the world. Later in his career he wrote several books addressed to lay readers, among them *Why Buildings Stand Up* (1982) and *Why Buildings Fall Down* (1992). He died in 1997.[57]

Hans H. Bleich, an Austrian of Jewish parentage trained as a civil engineer in Vienna, left Europe in 1938 for New York City. He became an assistant professor of civil engineering in 1946; in 1954 he became director of Columbia's Guggenheim Institute of Air Flight Structures. He helped design the Palomar and Mount Wilson Observatories. He was the winner of the von Karman Medal awarded by the ASCE in 1973 and was elected to the NAE five years later. Bleich remained at Columbia until retiring in 1977. He died in 1985.[58]

Bruno A. Boley came as a boy from Trieste, Italy, in the late 1930s to New York City, where he took his undergraduate degree from CCNY and

ScD degree in aeronautical engineering from Brooklyn Polytechnic Institute. He then worked briefly in industry before taking a teaching position at Ohio State. In 1952 he came to Columbia as a professor of civil engineering. In 1960 he wrote, with his colleague Jerome H. Weiner, *Theory of Thermal Stresses*. Boley left Columbia in 1968 to become the founding chair of the theoretical and applied mechanics department at Cornell University. He later moved on to Northwestern where he was dean of the engineering school. Upon retiring from Northwestern in 1992, he returned to Columbia as professor of civil engineering.[59]

Alfred M. Freudenthal was a Pole of Jewish parentage and trained in Czechoslovakia (PhD Prague, 1930) who emigrated first to Palestine in 1935 and then in 1947 to the United States. In 1949 he became professor of civil engineering at Columbia and in 1962 director of the Institute for the Study of Fatigue and Structural Reliability. In 1969, distressed by campus turmoil at Columbia, Freudenthal and his institute moved to George Washington University. His work at Columbia and GWU earned him the accolade "the father of structural reliability." The ASCE Alfred M. Freudenthal Medal, first awarded in 1976, recognizes "distinguished achievement in safety and reliability studies applicable to any branch of civil engineering." He died in 1977.[60]

Richard Skalak, a Columbia product (BA 1943, BS 1946, PhD in civil engineering 1954) was appointed an assistant professor of civil engineering in 1954, having taught for several years as an instructor. Early in his career he focused on traditional engineering topics like fluid turbulence and research associated with finding and extracting oil. In Sweden on sabbatical in 1966, he directed his interest in fluid flows to study how red blood cells flow in human tissue. The resulting paper, "Deformation of Red Blood Cells in Capillaries," the first study to trace blood movement in quantitatively measurable terms, appeared in the journal *Science* in 1969. This and subsequent collaborations with Columbia medical school colleague Shu Chien established them as the pioneers in the field of biomedical engineering. Upon their retirement from Columbia in 1987, Skalak and Chien continued their productive collaboration at the University of California, San Diego. Skalak died in 1997.[61]

To these thirteen elected members of the National Academy of Engineering might be added two others of the sixty-eight-member 1955 SEAS faculty who achieved eminence in their respective engineering fields but were not elected to NAE. David B. Hertz (EE 1939, CU PhD 1953), was a member of the industrial engineering department specializing in the then new field of operations research from 1948 to 1959, when he left academic life for Wall Street to pioneer in the use of Monte Carlo methods in finance and become one of the world's first "quants." He later became a principal member of the consulting firms McKinsey and Co. and Arthur Anderson before returning to academic life at the University of Miami. He died in 2011. John Ralph Ragazzini (CCNY 1932, Columbia 1941) became an assistant professor of electrical engineering in 1941 and an early leader in the field of control theory. He was a mentor of Eliahu Jury, Rudolf Kalman, and Lotfi Zadeh, all of whom have since been Egleston Medal recipients. Ragazzini left Columbia in 1958 to become dean of the NYU engineering school. In 1979, the American Automatic Control Council named the John R. Ragazzini Award and made him the first recipient of the award. He died in 1988.[62]

How, besides their greater national recognition, did these fifteen postwar Columbia engineering notables on the 1955 faculty differ from their fifty-three colleagues? On two scores they were alike. The first was the incidence of receiving their academic training at Columbia. Gaden, Mindlin, Kellogg, Hertz, Skalak, and Arbiter had all their degrees from Columbia, while only Drew and three of the European émigrés, Bleich, Salvadori, and Freud-enthal, had none of their training at Columbia—so much for the timeless universality of the dictum against faculty inbreeding. But there were also differences. The school's future NAEers were heavily concentrated in civil engineering, where they constituted half the department. Four of the six were émigrés, two of them Jewish, allowing the conclusion that no department in the university took more advantage than did civil engineering of the Nazi-induced European intellectual migration. Their presence reinforced an already present ethic of collective scholarly enterprise, a factor emphasized by Frank Dimaggio and Rene Testa, both graduate students in the 1950s, in calling the postwar period their department's "golden age." They are not alone in thinking so. In 2002 the Stanford structural engineer Jeremy Isenberg described Columbia civil engineering in the 1940s and 1950s as "the nation's most prominent center for engineering mechanics."[63]

And yet the cohort of future NAEers differed from their colleagues in another way: they were much more likely to leave Columbia in mid-career. More than half did so. Four departures could be explained by offers involving greater academic or administrative responsibilities: Ragazzini to NYU as its engineering school dean; Zadeh to Berkeley to start up its computer science department; Boley to Cornell to chair a new department; and Drew back to MIT to chair its much larger chemical engineering department. Hertz left for Wall Street. Arbiter took early retirement to return to industry. But Freudenthal's departure to George Washington University in 1969 turned in part on his frustration with the Columbia administration's response to student disruptions, and Elmer Gaden's leaving for an administrative post at the University of Vermont School of Engineering followed on his complaint to university officials about the absence of administrative leadership at the engineering school. Neither could be characterized as other than lateral career moves.

Of the six future NAEers who stayed on at Columbia, Salvadori, Bleich, and Freudenstein were European émigrés who became committed New Yorkers by adoption and were not about to be lured elsewhere. This left Harris, much of whose acoustical work centered in New York, Skalak, who continued his career after retiring from Columbia at UC San Diego, and Mindlin, the born and bred New Yorker, who despite outside offers assured anxious department colleagues he would stay at Columbia as long as Jewell Garrelts chaired his department. Nonetheless, the fact that eight of the school's fifteen leading faculty left Columbia for elsewhere must have been, for both their equally luminous colleagues and other faculty who stayed behind, dispiriting.[64]

These departures were by no means limited to the engineering school or to its stars. Columbia provost Jacques Barzun later described the postwar years at Columbia as "the era of the packed suitcase." Once one of the more aggressive "raiders," Columbia in the early 1960s had itself become regularly subjected to systematic raiding, not only from its traditional competitors, but from the rapidly expanding units of the State University of New York and the California system buying up name-brand faculty in bunches. Other parts of Columbia, the area studies programs and the humanities departments, had more success in fending off raids because the president and provost cared more about them than they did the engineering school. In 1962 a special budget line was created for departments in arts and sciences to permit department chairs to make counteroffers to hold their most valued faculty. No such fund seems to have been provided the engineering school. Whether pulled by enticing offers or pushed by local irritations, the regular and seemingly accelerating departure of Columbia's top engineers throughout this period contributed significantly to the school's subsequent reputational slide.[65]

Another contributing factor was scale. Although the Columbia engineering school faculty doubled its size between 1950 and 1965, it failed to match the growth of its already substantially larger East Coast competition, MIT and Cornell, and even less the growth among the state-funded University of California, Berkeley, Michigan, and Illinois. Nor did it keep pace with the newly competitive Stanford and Georgia Tech, whose engineering faculties grew in the postwar years at a rate easily twice that of Columbia. The result was that Columbia's engineering school found itself in the mid-1960s with a faculty, relative to its peers, smaller than it had been in the 1930s. True, Harvard, Yale, Princeton, Dartmouth, and Brown experienced equally modest growth and in some instances shrinkages. This provided small satisfaction to non engineering Columbians who thought they ought to have either a world class engineering school or none at all.[66]

"NEARER TO GOD"

Whenever one of my students came to me with a scientific project, I asked only one question, "Will it bring you nearer to God?"

—I. I. RABI, PROFESSOR OF PHYSICS[67]

The engineering school suffered from an intense inferiority complex.

—MORTON FRIEDMAN, ON HIS EARLY YEARS AT SEAS[68]

Space constraints, faculty raiding, restrictions on student recruitment, the movement of jobs away from New York, all these liabilities made it diffi- cult for Columbia's postwar engineers to keep up with their faster growing and more rustic peers. While other Columbia faculty and students operated under the same conditions, one that did not obtain elsewhere at Columbia but was peculiarly felt by its engineering school was the persistent skepti- cism as to whether it even belonged at Columbia.

First articulated by Columbia political scientist John W. Burgess in 1884, reinforced by the Columbia psychologist James McKeen Cattell and the Columbia philosopher John Dewey in the early 1900s, and acquiesced to by engineering alumnus Gano Dunn, Columbia's technology-averse academic ethos had survived World War II intact. Postwar articulators of this posi- tion included physicist I. I. Rabi, literary critic Lionel Trilling, and cultural historian Jacques Barzun. All three graced the front cover of *Time* at one time or another during the postwar years.[69]

The case for Rabi as a technology skeptic turns more on his espousing a privileged place of basic science in the university than as a physicist, where he did some of his best work as an experimentalist with funding from very results-oriented patrons. During World War II his first ques- tion when reviewing a proposal by his MIT-based colleagues was wholly operational: "Will it kill more Germans?" Yet once the war was over, even with his Radiation Lab back at Columbia now funded by the Army Signal Corps, Rabi defined the kinds of research acceptable as characterized by "impracticality" and "frankness," which he contrasted with work carried on at industrial labs and by applied scientists generally. Rabi's authorized biographer John Rigden quotes him reminiscing: "Whenever one of my students came to me with a scientific project, I asked only one question. 'Will it bring you nearer to God?'"[70]

Rabi was not only a Nobel laureate physicist, he was the first faculty member to be awarded Columbia's highest faculty honor, a university pro- fessorship, and for forty years epitomized Columbia science. Whatever disregard he expressed toward science not driven by "impracticality" but by commercial ends, a disregard shared by fellow Nobel laureates Charles Townes and Polykarp Kusch, carried over to nonscience colleagues.[71]

Two of Rabi's Columbia colleagues, both in their time university professors, were Lionel Trilling and Jacques Barzun. They were friends, taught a famous seminar together, oversaw a commercially successful book club, and exemplified the humanities for the postwar generation of Columbians. Both found occasions to pronounce on the place of science in the academy and in Western culture. Insofar as they saw that place a modest one, their views had resonance on Morningside Heights.[72]

One of Barzun's early forays away from his first scholarly base in European intellectual history into contemporary American culture was *The House of Intellect* (1959). As he later acknowledged of the book's argument, "some may hastily infer that I am 'against science.'" This acknowledgment occurs in the introduction of his subsequent *Science: The Glorious Entertainment* (1964). In the introduction, he wrote:

> Science itself, through its increased prosperity and fact-breeding specialism, has contributed to the disturbance and distress [of the past thirty-five years], while the technology of automation, nerve gas, germ warfare, and atomic destruction have driven home the lesson of human helplessness directly proportional to human "control over nature."[73]

Perhaps not surprisingly, no published review of *Science: The Glorious Entertainment* by a Columbia engineer has turned up. Thirty years later, however, Wallace S. Broecker, then and since Columbia professor of geochemistry, recalled the book's impact on himself and other Columbia scientists. In doing so, he also takes a gratuitous blame-shifting shot at the expense of engineering:

> *Science: the Glorious Entertainment* really annoyed me. Because basically in that book he blamed everything on science. You know, row housing, every bad thing about society he sort of attributed to scientists. Well, you know, that's hardly the case. You know there were people making money off of using engineering, basically, but not really science.[74]

It should be said that Barzun is equally critical of those engaged in technology (he preferred the term "techne") and explicitly rejected one of the tenets of technophobia when he wrote, "It is not true, for example, that technology is the offspring of science." In this he differs from the sociologist Daniel Bell, for more than a decade Barzun's Columbia colleague, for whom technology was merely an epiphenomenon of science. From Bell's *The Coming of the Post-Industrial Society* (1973): "The computer would not exist without the work in solid-state physics initiated forty years earlier by Felix Bloch. The laser came directly from I. I. Rabi's research thirty years ago on molecular nuclear beams." For all of Barzun's relative even-handedness, he stopped short of turning a sympathetic eye toward what was, after all, one of his provostial constituencies.[75]

In the same year *Science: The Glorious Entertainment* appeared, Barzun's longtime friend and Columbia colleague, the literary critic Lionel Trilling, took the occasion of the Cambridge professor of literature F. R. Leavis's entry into the already crowded controversy arising from C. P. Snow's Rede Lectures, *The Two Cultures and the Scientific Revolution*, to weigh in. Trilling began his 1963 essay, "The Leavis-Snow Controversy," by chastising Leavis for his "attack of unexampled ferocity upon the doctrine and the author of *The Two Cultures*." Trilling thereafter dismissed Snow's central complaint that "it is the traditional culture ['literature'], to an extent remarkably little diminished by the emergence of the scientific one, which manages the Western world." For this and similar statements, Trilling declared *The Two Cultures* a book "which is mistaken in a very large way indeed."[76]

Just before pronouncing this judgment, Trilling, seeming about to let Snow off the hook, ended up confirming his Columbia anti-technology bona fides:

He [Snow] does say that scientists need to be "trained not only in scientific but in human terms," but he does not say how. Scientists— but eventually one begins to wonder if they are really scientists and not advanced technologists and engineers.

Not really scientists, only "advanced technologists and engineers." Little wonder then that given the intellectual climate at Columbia at the time of his arrival in 1956, the applied mathematician and civil engineer Morton Friedman said of his fellow engineers: "We suffered from an intense inferiority complex."[77]

ENGINEERING AND THE RANKINGS GAME

Departmental rankings and the institutional rankings derived from them became an unavoidable fact of life in the mid-1960s. Since then every president or dean has had occasion to grouse about this or that metric that puts his or her university at a disadvantage. Of all universities, Columbia has the least excuse to complain. It was, after all, a Columbia professor, the psychologist James McKeen Cattell, who started it all in his 1903 edition of *American Men of Science*. Cattell thereafter published his rankings of American scientists and science departments over the next four decades, but, in keeping with the prevailing technology-averse ethos of the scientific community, resolutely refused to include engineers and engineering departments in his rankings.[78]

Other researchers, notably Raymond Hughes in 1925 and Howard Keniston at the University of Pennsylvania in 1948, extended rankings to departments in the humanities and social sciences, while relevant professional societies took up the periodic ranking of law and medical schools. It was only with the 1966 publication of *An Assessment of Quality in Graduate Education* by the American Council of Education, with the results of a 1964 survey of four thousand faculty members in thirty disciplines at 106 major institutions by the Princeton economist Allan M. Cartter, that survey efforts acquired sufficient statistical reliability to require attention.[79]

The Cartter survey rated four engineering disciplines—chemical, civil, electrical, and mechanical engineering—by the same two criteria used to rate departments in arts and sciences: "the quality of graduate faculty" and "the effectiveness of doctoral program." Three of Columbia's four ranked engineering departments placed among the top ten in faculty quality: civil engineering ranked eighth; electrical and mechanical engineering ranked tenth. Chemical engineering ranked out of the top twenty departments.[80]

Looking back on these rankings in 1970, the dependably acerbic Columbia provost Polykarp Kusch characterized them as "very respectable." To others in 1966, however, they were troubling on at least two counts. The first was the unavoidable conclusion that reputational standing and department size were highly correlated, if not determinative. With the exception of Caltech, sui generis in many ways, small engineering schools like Columbia's could not hope to compete for ratings with larger schools like MIT, Stanford, Wisconsin, the University of California, Berkeley, and the University of Illinois, all of which had two or more of their engineering departments in the top five in the Cartter rankings.[81]

A second problem was the fact that the three Columbia engineering departments ranked in the top ten for "quality of faculty" did less well on the "quality of graduate program," with only civil engineering making it into the top fifteen. Insofar as "Faculty Quality" was regarded as a lagging indicator, reflective of the past accomplishments of individual faculty, while "Program Effectiveness," a leading indicator speaking to the range of a department's curricular offerings and the promise of its recent graduates, Columbia's relative weakness in the latter rankings spelled trouble ahead.[82]

Columbia's major science departments (biological sciences excepted) consistently numbered among the top ten, with the geology department (aka Lamont-Doherty Earth Observatory) fourth, physics and chemistry seventh, and mathematics ninth.[83] Similarly high rankings were accorded Columbia's other professional schools. The law, medical, business, and architecture schools, as well as the affiliated Teachers College, all ranked in the top five in surveys of the period. Columbia's School of Journalism often ranked first. For technology-averse Columbians the question remained: If the university was not to have a top-ranked engineering school, should it have one at all?[84]

"NOBODY GAVE A DAMN": FORFEITING COMPUTER SCIENCE

The story of computing at Columbia in the postwar period provides a classic instance of Finch's "missing the boat." It tells how space constraints, a science faculty's indifference to applied science, administrative neglect at the university level, and the absence of imaginative and assertive leadership in the engineering school combined to cause Columbia to forfeit its early prominence in the field of computers and computer science to more nimble competitors, chiefly MIT, Stanford, and Carnegie Mellon.[85]

The history of computing, defined initially as the history of constructing machines to process and manipulate numerical data, traces its American origins to the work of Herman Hollerith, an 1879 School of Mines graduate. In 1890 Hollerith was awarded a Columbia PhD in political science for a dissertation on immigration that made use of a mechanical counter the author invented to record vital statistics of immigrants at Ellis Island as part of the 1890 federal census. Six years later Hollerith founded the Tabulating Machine Company. TMC's principal product was the "Hollerith Machine," which recorded data by holes punched in cards that could then be electromagnetically sorted and tabulated. In 1924 Thomas J. Watson, Sr., acquired Hollerith's Tabulating Machine Company as part of International Business Machines Company. In the early years, IBM sold its tabulating machines under the label "Hollerith Machines."[86]

In 1928, Watson, ever on the lookout for new ways to market his punch-card machinery, approached Columbia with an offer to install his latest tabulator and sorter in Hamilton Hall for the University's Bureau of Collegiate Educational Research. Bureau director Benjamin Wood promptly put the equipment to use sorting examination papers and tabulating the results of psychological tests carried out on students. In 1929 the Bureau of Collegiate Educational Research morphed into the Columbia University Statistical Bureau and migrated to Pupin Hall, where Wood had persuaded Watson to have IBM build and install a Special Difference Calculator (SDC). In addition to Wood's statistical needs, the SDC was used by Columbia astronomy professor Wallace J. Eckert to interpolate astronomical tables. When IBM developed commercial versions of the SDC, they were internationally marketed as "Columbia Machines."[87]

In 1934 Watson, who had skipped college but was now on social terms with Butler and several trustees, was elected a life trustee of Columbia. He quickly became one of the board's dominant members, even as he arranged for IBM's research and development activities to be based at Columbia. In 1937 the Pupin astronomical lab was renamed the Thomas J. Watson Astronomical Computing Bureau. Three years later, just before leaving Columbia to become wartime director of the *Nautical Almanac* at Naval Observatory in Washington, Eckert, then on the IBM payroll, published *Punched Card Methods in Scientific Computations*, considered the first textbook on computers.[88]

Frank da Cruz, who has written the most detailed story of Columbia computing, is surely correct in insisting that "astronomy, and not engineering, was at the center of the development of modern computing [at Columbia]." During World War II Columbia's computers were used to calculate weapons trajectories and to compile statistics on propaganda projects for the Bureau of Radio Research, directed by mathematical sociologist Paul Lazarsfeld. Eckert returned to Columbia as professor of astronomy and director of the new Thomas J. Watson Scientific Computing Center at Columbia University, located in a brownstone on West 116th and later in its present building on 115th Street.[89]

News of the imminent opening of the Watson Computing Lab prompted Dean Finch to alert Acting President Frank Fackenthal to the school's plans to secure a Marchant Automatic Calculating Machine for its electrical engineers. He also expressed the school's hopes for a role in the operations of the Watson Lab. "You can count on our not forgetting about Engineering," Fackenthal assured Finch. The following spring the engineering school mounted the first graduate-level course in computer science offered at any university, Engineering 281, taught by the mathematical astronomer (and IBM staffer) Herbert Grosch. Meanwhile, Watson lab staff designed and built the IBM Selective Sequence Electronic Calculator (SSEC), one of the first large-scale stored-program electronic computers. On the Watson staff at the time was John Backus, who in 1954 designed FORTRAN, the first high-level machine-independent programming language.[90]

For all this activity, Eckert, now wholly on the IBM payroll, was one of the few Columbia scientists in the late 1940s and 1950s to reckon fully with the revolutionary potential of computing. An indifferent teacher, immune to the allure of empire-building and a loner, Eckert unfortunately lacked the requisites of an evangelist for this "new new thing." Consequently, academic computing at Columbia in the 1950s fell through the interschool administrative cracks, with neither the Faculty of Pure Science nor the Faculty of Applied Science giving or taking responsibility for it. In 1957, when the National Science Foundation offered Columbia a $200,000 grant to install the new IBM 704 computer on campus, the offer was turned down by the central administration when it refused to commit the space to house it. The IBM 407 then went to the more accommodating Harvard.[91]

A year later, the horse well out of the barn, Dean of Graduate Faculties Ralph Halford convened a committee to examine Columbia computing. Its membership consisted of Eckert, mathematician Samuel Eilenberg, and the physicists Richard Darwin and Polykarp Kusch. By then the engineering school offered the occasional course in computing, and the civil engineer Mario Salvadori volunteered his services to the committee, but nobody from the engineering school was appointed. This may have been as the members wished it. Of them, Eilenberg showed throughout his distinguished career what some saw as a lively disdain for applied mathematics and for applied science generally. "There is pure mathematics," a colleague remembers him pronouncing, "and there is bad mathematics." Kusch held Columbia's engineers in similarly low regard, reportedly mocking a new member of the Krumb School of Mines, Colin Harris, as Columbia's "professor of steel." Not surprisingly, the committee report evinced no great concern that Columbia's early lead in computational science had all but vanished.[92]

Meanwhile, in 1956, Thomas Watson, Sr., was succeeded as IBM CEO by his son, Thomas Watson, Jr. An alumnus of Brown and a generous benefactor of his alma mater, the younger Watson had no personal ties to Columbia. Although some IBM staff lingered on at 115th Street into the

1960s, by then Big Blue's major R & D effort had relocated to Tarrytown, New York, and points west.[93]

It was just at this point that Edward Leonard and Joseph Traub came to Columbia, Leonard as a graduate student from the University of Pennsylvania and Traub a beginning graduate student in applied mathematics. Both recall being struck by the almost unlimited access they had to Columbia's still state-of-the art computers in the Watson Lab, which is also to say by the lack of use made of them by Columbia faculty and other graduate students. During Traub's three years at Columbia (1956–1959), the five members of the Committee of Applied Mathematics supervised one or two dissertations a year, with Traub's dissertation in 1959 on optimal iteration theory virtually alone in making extensive use of the university's hardware for computational analysis. Part of the problem, as Traub later explained, was that Columbia in the late 1950s was "very strong in physics and very strong in mathematics, but so weak in applied mathematics and numerical analysis." But part of the problem was also that when it came to the field of computing where Columbia had once enjoyed considerable standing, "Nobody seemed to give a damn."[94]

Not until twenty years later when no longer able to ignore the reality that a university without a significant presence in computer science could not compete with universities like MIT, Stanford, and Carnegie Mellon, Columbia moved to reclaim its place in computer science, having to do so pretty much from scratch.

POSTWAR ENGINEERING STUDENTS

The restrictive admissions policies of the interwar period had limited the engineering school to providing training to only a small percentage of the engineering-bound graduates of New York City's top high schools. Upwards of half of these graduates were Jewish. Whether denied admission or (as in the case of Joe Traub) not applying because of Columbia's anti-Semitic reputation, they went instead to CCNY, NYU, Cooper

Union, Brooklyn Polytech, or Manhattan College. Many then went on to distinguished careers in engineering and engineering education, including a dozen graduates of local engineering programs who after the war took up places on the Columbia engineering faculty.[95]

Interviews with Jewish alumni who attended the engineering school in the 1930s yielded no recollections of overt anti-Semitism, either at the administrative level or among the school's non-Jewish faculty and students. Still, Columbia's reputation for discriminating against Jewish applicants persisted into the 1950s even after the discriminatory policies had been jettisoned in the immediate wake of World War II. What followed, as recalled in the 1970s by Columbia historian Fritz Stern (CC 1946), was a brief fifteen-year window when Columbia College (and thus the engineering school with the College still its primary portal) had abandoned the restrictive policies, and the early 1960s, when MIT, Stanford, Cornell, and others began aggressively recruiting New York City public school graduates, when the College and the engineering school enjoyed privileged access to the city's scientifically gifted talent pool which both had earlier spurned.[96]

Religious discrimination a thing of the past, the intrainstitutional provision by which the engineering school was limited to recruiting students from among those who had completed either two or three years of pre-engineering at Columbia College remained in force. (Transfers from thirty-odd "cooperating colleges" accounted for only a handful of incoming engineering students.) Not until the fall of 1959, after a forty-seven-year hiatus, did the engineering school again admit newly minted high school graduates. One factor prompting the change of heart among Columbia administrators was that the prohibition against admitting freshmen caused Columbia to miss out on the national surge of engineering enrollments following on the Russian launch of Sputnik in the fall of 1957. The entering freshman class in 1959 consisted of seventy-five students (including two women), who enrolled in either the new four-year program leading to a BS in engineering, or in the older five-year program leading to a professional degree.[97]

While allowing the engineering school to admit first-year students made sense to both the school and the central administration, doing so did not sit well with Columbia College. Dean John Palfrey's first public reaction to the decision, with its intention of increasing engineering enrollments, expressed concern about its implications for the College. Five months after the first class of four-year engineers arrived on campus, Palfrey returned to the issue in a talk that secured a *Spectator* headline, "Dean Fears Engineers Strain College Facilities." Referring to the 1956 vote of the College faculty against expanding the College, he declared the new engineering program "does not produce the consequences for Columbia College that its faculty sought to avoid in considering its own size." He pointed specifically to the problems larger classes of entering engineering students would place upon the chronically understaffed core courses supplied by the College. This refrain—the engineering school was draining scarce resources rightfully the College's—would be one that Palfrey's successors regularly invoked over the next thirty years.[98]

Even with a continued flow of transfer students from the College under the traditional 3/2 plan, total undergraduate enrollments in SEAS in the early 1960s remained in the 500–600 range, with entering classes averaging around 150. Mudd, which opened in 1962, was built to accommodate three times that number. More troubling was the fact that, having missed out on the immediate post-Sputnik surge in enrollments of the late 1950s, SEAS became a full participant in the mid-1960s nationwide undergraduate slump in engineering enrollments that followed. The Class of 1964 lost 50 of its original 124 students, some transferring to the College and others leaving Columbia. The school, unlike nearly all other major engineering programs, where undergraduate enrollments were a multiple of graduate enrollments, enrolled more graduate students, most of them part-time and underwritten by their employers. However agreeable it may have been for faculty who preferred teaching graduate students, an engineering school relying heavily on graduate enrollments did so at its financial peril.[99]

The good news was that occasional grumblings emanating from Hamilton Hall, disappointing national rankings, and flat enrollments had little effect on the quality of the academic program the engineering school provided in the postwar era. Early on, a majority of the undergraduate engineers came as veterans, funded through the generous terms of the GI Bill. Older than the typical undergraduate, in many cases they were married. Still, many of them took part in intercollegiate sports and other campus activities. These veterans later became some of the school's most loyal alumni.[100]

Most engineering graduating classes after 1945 included one or two women, but never more than five well into the 1960s. More numerous were the engineering undergraduates who beginning in 1946 came to Columbia on Navy ROTC scholarships, which secured them regular commissions upon graduation. Still other engineers signed on as NROTC contract midshipmen as an alternative to the draft and to secure reserve commissions upon graduation. (This mutually advantageous arrangement with the Navy was a casualty of the disruptions in 1968, but recently restored.) Engineering undergraduates from New York also benefited from the then generous terms of the New York State Regents fellowship program, which into the late 1960s covered upwards of two-thirds of their tuition.[101]

In the days before student evaluations, anything said about the quality of classroom teaching is necessarily anecdotal. Three data points provide some insight: Gloria Brooks Reinish—an undergraduate from 1943 to 1945, an MS student from 1948 to 1949, and a PhD student in the 1970s—recalls the quality of the teaching as being better during her first two exposures than her third. A letter to *Pulse* in 1964 complained about four assistant professors in civil engineering being denied tenure, one of whom the letter writer described as "maybe the best teacher in the school." When asked in a 2009 interview whether he had any great classes at Columbia as a graduate student in the Faculty of Pure Science in the 1950s, Columbia geochemist Wallace S. Broecker, cited one taught by an engineering professor:

Well, one thing I had—I took differential equations at Columbia from Mario Salvadori. There were a hundred and twenty-five people in a room—where about eighty of them were smoking—in the evening for two hours. It was an absolutely wonderful course. Nobody ever dozed at all. This guy was just a marvelous Italian. Never had to read a book, nothing. I understood exactly after I could do the problems. I never had to think about it because he was such a good teacher.[102]

Another factor that contributed to social cohesion and school loyalty among an otherwise diverse engineering student body was Camp Columbia. Located outside Litchfield, Connecticut, and encompassing some 400 acres of farmland, the once working farm had been a Columbia institution since the 1890s, where engineering students acquired firsthand experience in the rudiments of surveying and construction site management. The football team also held their preseason practices there. During the years when the 3/3 or 3/2 plan was in effect, students transferring from the College to engineering school went to Camp Columbia as part of their rite of passage. After 1959, entering students were required to attend a four-week program there before starting classes in the fall.[103]

Concern about the future of Camp Columbia was raised in 1962 as a budgetary issue, which led the following year to the launching of *Pulse,* the engineering school's own student-run newspaper. *Pulse* staffers conducted surveys to assess community sentiment about continuing required attendance at Camp Columbia. Responding alumni strongly favored doing so, while faculty and students were divided. Students opposed pointed out that the four-week program interfered with summer employment. Others questioned the relevancy of the focus on surveying for students not going into civil engineering. On March 10, 1965, the *Pulse* editorial board came out in favor of closing Camp Columbia. Two weeks later *Pulse* reported the faculty's concurrence, by a close vote of 35 to 31 in favor of dropping the Camp requirement. Some alumni still regard this decision to have been a serious mistake.[104]

Although Camp Columbia was no longer, *Pulse* continued to be published by engineering students into the 1970s. Dean Hennessy described engineering students in 1965 as "The Quiet Revolutionaries," and intended it as a compliment; as to the adjectival ascription, *Pulse* editors and contributors provided ample contrary evidence.[105]

The opening of the Seeley Wintersmith Mudd building in 1962 and the Engineering Terrace two years later provided faculty with much needed office and laboratory space, but also students with what in the early 1960s were regarded as "state of the art" classrooms and extraclassroom amenities in buildings that for the moment were exclusively theirs. Both provided Columbia engineers, after six decades of scattered residence, with a home of their own. Never mind that Mudd was tucked over in the northeast corner of the Morningside campus and that the building resembled the "box" that the other Columbia architectural scandal of the era, the Business School's Uris Hall "toaster," came in.[106]

6

BOTTOMING OUT 1965–1975

The situation of the [Engineering] School is clearly intolerable and cannot long continue.

—VICE PRESIDENT POLYKARP KUSCH TO INCOMING
PRESIDENT WILLIAM MCGILL, JULY 8, 1970[1]

Ever since the Engineering School has existed, the College has been dictating the policies of the University. Our School has always had to play second fiddle to College administration. . . . We urge the School's administration to show more courage to face such crises. Our School has been pushed around enough and it is now time for us to fight.

—EDITORIAL "ENOUGH IS ENOUGH," *PULSE*, SEPTEMBER 27, 1973[2]

Your School of Engineering at Columbia has lost its way. . . . No one at MIT takes the Columbia engineering school seriously. No one knows who the Dean is.

—GORDON BROWN, DEAN EMERITUS OF MIT, OCTOBER 4, 1973[3]

A DECADE-LONG CAMPUS conflict began at Columbia in the spring of 1965. Protests against American military involvement in Southeast Asia and parallel actions against the university's role as Morningside's dominant landlord became regular events. Demonstrations escalated as the war in Vietnam intensified and town–gown relationships deteriorated. The most disruptive events occurred in the spring of 1968, when protesters occupied five campus buildings for a week and precipitated a violent police action that led to the arrest of 705 Columbia students. These disturbances forced the shutdown of the university four weeks before the end of the semester, produced competing commencement ceremonies, and brought about the resignation of President Kirk.[4]

Student disruptions remained a feature of campus life on Morningside Heights after 1968, but by the early 1970s the fundamental character of Columbia's ongoing crisis shifted from one driven by competing ideologies ("liberationist/revolutionary" vs. "institutionalist/preservationist") to one of financial exigencies of "what goes" versus "what stays?" This financial crisis put all of Columbia's schools, departments, and programs to the test of justifying their continuance. Only three arguments served:

—financial sufficiency—"We pay our way";
—quality—"We're too good to eliminate";
—institutional centrality—"There's no Columbia without us."

No longer would "we do no harm" or "we are not a bother" cut it. Under this new dispensation the case for the School of Engineering and Applied Science in this the most Darwinian decade of Columbia's history was far from ensured; its survival hung in a decade-long balance.[5]

KIRK AND THE ENGINEERS

What Jacques Barzun called "the long interregnum that put Columbia at a disadvantage in the early fifties" encompassed the last years of the Butler presidency, the Fackenthal acting presidency, and the absentee presidency of Dwight Eisenhower. It was thought to have ended in 1953 with the installation of Grayson Kirk as the sixteenth president of Columbia University. Kirk had been provost under Eisenhower for four years and knew the ways of Washington (having served in the wartime State Department) and the major private foundations (especially in securing support for Columbia's many area studies programs). True to expectations, the first decade of his presidency was widely viewed to be a success.

Yet by the early 1960s Columbia found itself struggling. It had become increasingly dependent upon federal support to expand its activities and balance its books. By 1963, 45 percent of the annual income of the university came from federal contracts. The dangers inherent in such reliance upon "soft money" led the trustees in 1964 to authorize the first All-University Capital Campaign, with a launch set for the spring of 1966. The campaign's goal of $200 million was the highest ever sought by any American university. Kirk, at sixty-six and in his twelfth year as president, reluctantly took up the begging cup.[6]

Another problem, exacerbated by the about-to-be-launched capital campaign, was the growing incidence of campus disruptions. These began in the early 1960s, instigated by overlapping groups of students, for the most part undergraduates well removed from the "silent generation" who preceded them. These students, several of them self-described "red diaper babies," viewed the university as culpable for all manner of social ills, from American military involvement in Southeast Asia, to the ongoing mistreatment of blacks in the rural South and urban North, to the deterioration of the Columbia neighborhood. In the spring of 1967 students affiliated with the Students for a Democratic Society (SDS) began a campaign of openly challenging university officials, many of whom were even more ill-suited to the thankless task of dealing with often ill-mannered and righteous undergraduates than was Kirk, who labored under the personal challenge of being unable to speak informally without stuttering. Sensing that the president wished for above all else a quiet campus at the launch of a fundraising drive, radical students proceeded to deny him that.[7]

Another problem facing Columbia's president in the early 1960s was the dearth of effective leadership at several professional schools. Kirk and his pugnacious business school dean, Courtney C. Brown, regularly crossed swords, often to the president's dismay. But the engineering school, since 1961 the School of Engineering and Applied Science (SEAS), was the most striking instance of administrative dysfunctionality. Dunning's relegation to the Mudd basement in the spring of 1964 left the school's direction in the hands of Executive Dean Wesley Hennessy. A competent and well-liked staffer during his quarter century at SEAS, Hennessy was ill equipped to take the school in new directions in the best of times, which these were not.[8]

THE POLITICS OF COLUMBIA ENGINEERS

Convention has it that engineers, white-collar professionals with ties to industry and the marketplace, make poor bomb throwers. True enough with Columbia's engineers. "Butler tells me that all of the trouble is in those departments having to do with things in the abstract," Trustee Benjamin B. Lawrence (EM 1878) wrote in 1917, to Trustee Chair William Barclay Parsons (EM 1882). Parsons, away on war duty in Paris, had inquired regarding the Columbia faculty's reaction to President Butler's recent firings of three professors for their antiwar activities. "The Schools of Engineering have supported him, and the Schools of Law and Medicine are all right," Lawrence reported. "The agitation comes from teachers in Psychology, History, and Philosophy, as might naturally be expected."[9]

So too it must have been in 1928, when the engineering faculty backed the "Great Engineer" Herbert Hoover for the presidency over his Democratic challenger, New Yorker Al Smith. Deans Barker and Finch were both active Republicans. In 1952, when competing campus petitions were circulated supporting the Democratic presidential nominee, Illinois governor Adlai Stevenson, and the Republican nominee, Dwight Eisenhower, Columbia's engineers mostly liked Ike. Dunning certainly did.[10]

Some of the engineering school's immunity to radical proselytizing may turn on its history of good relations with the military, especially the Navy. Many Columbia engineering faculty had served in uniform during World War II while others accepted special wartime assignments or remained on campus instructing naval midshipmen. In the immediate postwar period, when the federal contracts system came into being, engineers expressed fewer misgivings about federal funding of university research, most of it provided by military agencies, than did President Eisenhower or faculty in the humanities.[11]

Antiwar protests began on the Columbia campus with the picketing of an NROTC commissioning ceremony in June 1965. That summer a chapter of Students for a Democratic Society (SDS) was established at Columbia, its elected leaders subsequently orchestrating a number of escalating protests on campus. At the peak of its influence in the spring of 1968, SDS had some one hundred members, with possibly another two hundred "fellow travelers." Membership in other student groups critical of university policies may have doubled the total number of identifiable left-activist undergraduates to around five hundred, out of a total student population of seventy-five hundred in the four undergraduate schools. Most came from Columbia College but at SEAS the ranks of protesting students was likely limited to single digits.[12]

Among an engineering faculty of one hundred members in 1968, Professor of Industrial Engineering Seymour Melman and Professor of Mechanical Engineering John Englund played identifiably active roles in the antiwar movement. They opposed Columbia's participation in the Institute for Defense Analysis, a forum where leading universities and government agencies funding military research discussed issues of mutual interest. Recently retired, Professor of Mechanical Engineering Victor Paschkis, a Quaker-influenced pacifist who fourteen years earlier had funded the Society for Social Responsibility in Science, was another politically active Columbia engineer. Civil Engineering Professor Mario Salvadori, active in the civil rights movement and founder of a science school in Harlem, may have supported demonstrations mounted by black students on campus. Otherwise, SEAS faculty kept their politics to themselves.[13]

Noninvolvement remained the norm through the climactic campus events of April 23–30, 1968, when five academic buildings—Hamilton, Low, Avery, Fayerweather, and Mathematics—were occupied ("liberated" according to the occupiers) for the better part of a week by some fifteen hundred Columbia students and individuals from the neighborhood. Seeley Mudd and the Engineering Terrace escaped occupation, explained by some because of precautionary measures taken by Dean Hennessy. Others doubted they were ever at risk. Mark Rudd, Lewis Cole, and other College SDS stalwarts had likely never stepped foot in either building.[14]

The police action in the early morning of April 30, 1968, that cleared the occupied buildings resulted in the arrest of 705 occupiers, 508 of whom were Columbia students. Of these, 11 were from the engineering school, comprising 2 percent of those arrested and .65 percent of the engineering school student body.[15]

Other evidence supports the view of engineers sitting out the events of 1968. In the midst of the building occupations, a rump ad hoc faculty group sought to interject itself in an arbitrative role between what its leaders viewed as the overwrought and out-of-touch university administration and the intemperate but well-intentioned students. Its all-night open meetings attracted upwards of two hundred faculty, mostly from the junior ranks, for varying degrees of time. Among the fifty or so identifiable regular faculty participants, only Seymour Melman was from the engineering school. Opinion polls conducted by sociology graduate students and competing petition-signing campaigns carried out during and in the immediate aftermath of the occupations reveal two points: of all the university constituencies, engineers were the least supportive of the protesters; they were also the most supportive of a quick return to classes. In the days following the police action that turned much of the campus against the university administration for its handling of the protests, both President Kirk and Provost Truman congratulated Dean Hennessy for his school's support of the standing order. Alas, if life were fair, Columbia engineers should have escaped the financial and reputational fallout brought on by their revolution-bent fellow Columbians.[16]

COLLATERAL DAMAGE

Columbia's troubles did not end with the 1968 building occupations and police actions. Campus disruptions remained a fact of life for several more semesters and commencements remained boisterous affairs well into the 1970s. This was true of other campuses, including those with large and distinguished engineering programs. Protesting MIT students in 1969 forced the Institute to rename and disown its famed Instrumentation Laboratory and sever formal ties with the off-campus Lincoln Laboratory because of student opposition to secret research. That year Cornell experienced an occupation of Willard Straight Hall by a group of African American students, who subsequently left the building to be photographed carrying rifles and wearing bandoliers of ammunition. At Stanford in the spring of 1969 the Stanford Research Institute was occupied by protesting students who were forcibly removed by police using tear gas. At Wisconsin in 1970 the Army Math Research Center, housed in Sterling Hall, became the bomb target of protesting students, resulting in the death of a staff researcher.[17]

Even though Columbia was not alone in hosting violent student protests of the late 1960s and early 1970s it is the case that these protests more seriously threatened Columbia's viability than any other major university. Also clear is the fact that among Columbia's professional schools, the School of Engineering and Applied Science was among the most immediately at risk.[18]

Many factors combined to make the engineering school's situation particularly precarious as it entered the 1970s:

—the sharp drop in enrollments in the wake of 1968;

—an even sharper fall-off in alumni support;

—a further drop in departmental rankings;

—a caretaker dean in place;

—a new president prepared to undertake university-wide triage.

Enrollment in the engineering school had remained in the 1,500–1,700 range through most of the 1960s, with graduate students slightly outnumbering undergraduates. The impact of the spring 1968 protests on graduate enrollments was almost immediately felt when the school experienced a net loss of 250 graduate students registering for the fall term. The impact on undergraduate enrollments came the following year, the Class of 1972 having been admitted before the campus protests were national headlines. By the fall of 1970, declining undergraduate enrollments combined with stagnant graduate enrollments produced the school's lowest number in a decade. Admissions staffer Leonard DeFiore informed the SEAS faculty in the fall of 1970 that applications had dropped from 620 for the previous year to 500, with the likely outcome, Hennessy later acknowledged, "the smallest freshman class in the last five years." "I cannot emphasize too strongly," De Fiore added, "the seriousness of this situation."[19]

Alumni giving to the engineering school also fell sharply after 1968, even more than alumni giving to the College and the law school. The problem was further compounded at the engineering school by the ineffectiveness of its fundraising apparatus. In 1972, the school's fundraising operation cost forty-two cents for every dollar raised, about three times the ratio at the law school. A 1974 estimate by the Provost's Office had SEAS spending fifty cents for each dollar raised.[20]

Unfortunately for Columbia, the five-year follow-up to the 1964 Cartter survey of graduate faculties got underway with the circulation of questionnaires in the spring of 1969, when images of protesting students, occupied buildings, and police actions on the Columbia campus were freshly lodged in the memories of the 6,000 scholars and academic administrators charged with filling them out. But that was not the only problem. When published by the American Council of Education in 1970, *A Rating of Graduate Programs*, edited by Kenneth D. Roose and Charles J. Andersen, differed from the earlier *Cartter Assessment* in several aspects. Most importantly, whereas Cartter had only the 1925 impressionistic study by Raymond Hughes and the 1957 rankings compiled by Howard Keniston with which to compare its 1964 rankings, Roose and Andersen could compare their rankings with those of the methodologically compatible Cartter rankings of five years earlier. This allowed them to supplement the previous two rankings, "Quality of Faculty" and "Effectiveness of Doctoral Program," with a third—"Estimated Change in the Last Five Years."[21]

It was this third metric that put Columbia's situation in stark relief: according to this the university was experiencing a sharp and accelerating qualitative deterioration across the departmental boards. Of the fifteen ranked departments in the humanities and social sciences, only art history avoided losing ground, while two previously high-ranking departments, economics and political science, fell four and six places, respectively. Some of Columbia's science departments, physics and chemistry, held steady, but others, geology and psychology, tumbled. Of Columbia's ten ranked science departments, only astronomy was seen to have improved.[22]

Among engineering departments chemical engineering remained out of the top twenty, so relative change went unnoted. As for the other three previously ranked departments: civil engineering fell from eighth to tenth, electrical engineering from tenth to twenty-first, and mechanical engineering from tenth to seventeenth. This represented a collective loss of twenty places. By this same metric, Stanford's engineering departments climbed eight places, Cornell's four places, while MIT's, still the top ranked engineering school, fell two places. "The situation of the [Engineering] School," University Vice President Polykarp Kusch concluded upon reviewing the ACE rankings in July 1970, "is clearly intolerable and cannot long continue."[23]

Again, even in a period of retrenchment, comparisons of engineering's collective ranking with Columbia's other professional schools were both unavoidable and damning. Research underway at the time by the Columbia sociologist Peter Blau and his associate, Rebecca Margulies, published in 1973, yielded the following reputational ranking for several of Columbia's professional schools: journalism (1); social work (1); library science (2); education (2); law (3); theology (3); medicine (10). Blau and Margulies did not include SEAS in their list of leading engineering schools, the omission speaking volumes.[24]

Then there was the matter of administrative leadership. Dunning's retirement in January 1969 had provided Acting President Andrew W. Cordier a chance to install a new dean. He opted instead to appoint Hennessy, the previous "executive dean," as the school's seventh dean. Cordier's decision amounted to the third appointment of administrative convenience in the school's 105-year history.[25]

Columbia might well have had a difficult time in 1969 finding a promising outsider willing to come to a campus that had been the scene of student protests and police actions, and an engineering school with plummeting ratings. In addition, the acting president and trustees had more pressing matters to deal with than conducting a national search for a new engineering dean. Hennessy made the case himself for continuity in a spring 1968 interview with the *Columbia Spectator*: "I have been running the academic side of the School for five years." With ongoing administrative turnover a university-wide problem—three presidents and four provosts in three years—the case for continuity in at least one school likely struck Cordier

FIGURE 6.1 Wesley Hennessy (1914–1991), longtime engineering
school administrator and its eighth dean (1969–1975); known
for his attention to undergraduates.

Source: Columbia University Archives.

and the trustees as compelling. This was also the expressed view of the
engineering faculty. So the decision to stay with "Wes" was perhaps inevitable. Putting the best face on the appointment, Vice President Polykarp
Kusch said of Hennessy: "He is more of a realist than other deans and is
responsive to the inexorable pressure of economic reality." His appointment
ensured that any leadership-inspired change in the engineering school would
not be forthcoming.[26]

The fifth and most threatening problem facing the engineering school,
presidential skepticism as to its very necessity, requires a section of its own.

A NEW BROOM: MCGILL AND THE ENGINEERS

Grayson Kirk stayed on as president four months after his presidency had effectively ended with his decision, in April 1968, to call the New York City Police onto campus, leading to his formal departure. The dean of the School of International Affairs, Andrew W. Cordier, became acting president. During his twenty-six months as head of the university, Cordier, a seasoned and crisis-tested diplomat, restored a measure of peace to the campus, soothed trustee–faculty tensions and eased town–gown relations. He also exacerbated Columbia's financial problems by allowing budget deficits to continue unchecked. Cordier inherited a debt of $8 million; when he left two years later the debt stood at $34 million, and growing. Not since the early 1800s, when the College trustees went hat-in-hand to the vestrymen of Trinity Church for a life-saving bailout, had Columbia been so close to bankruptcy.[27]

The search for a permanent new president began upon Kirk's resignation in August 1968 and continued through 1969, and focused on a prospect who had once been at Columbia and left. In 1956 the newly arrived quantitative psychologist William J. ("Bill") McGill took a measure of his new home, comparing it with the institutions from whence he had come: "Harvard [and MIT] are about nature; Columbia is about books." Eight years later, he left Columbia for the chairmanship of the psychology department at the newly opened University of California, San Diego. His reasoning, he later recalled: "Morningside Heights was a thoroughly hostile environment. . . . Columbia was beginning to decline and I didn't want to go downhill simultaneously with Columbia's going down. . . . The future was in the west."[28]

Once in California, McGill quickly became caught up in its campus political struggles, both at San Diego and at Berkeley, where he attended meetings as an elected member of the university-wide faculty senate. In 1966 he was named chancellor of UC/SD and was immediately thrown into campus showdowns with disruptive students and radical faculty, including the formidable likes of Herbert Marcuse and Angela Davis. It was in these confrontations, which he later chronicled in *The Year of the Monkey*, that McGill "made his bones" and acquired the reputation of "one tough Mick."[29]

FIGURE 6.2 William J. McGill (1922–1997), psychologist and university administrator; sixteenth president of Columbia University (1970–1980); oversaw Columbia's recovery from the disruptions and financial problems of the 1960s.

Source: Columbia University Archives.

McGill's standing as a campus-conflict-hardened administrator, his years at Columbia, recommendations of Columbia faculty who knew him then, and his credentials as a New Yorker (he was born a few blocks from Columbia, grew up in the Bronx and attended Fordham) convinced the embattled trustees to offer him the presidency. In the spring of 1970 he accepted, later explaining, "because I was just so flattered to be asked." The fact that he was a nationally recognized scientist—the first since F. A. P. Barnard—appears to have entered into the trustees' deliberations not in the slightest.[30]

Before assuming the presidency in July 1970, McGill solicited from President Cordier's administrative team confidential assessments of various parts of the university. Eight days into his presidency he had in hand the "Report on Problems of the University" written by his inherited vice president and dean of faculties and Nobel laureate Polykarp Kusch. Kusch's report ranks among the most unvarnished internal assessments of a major university, with the sections on SEAS especially graphic. After citing the school's annual deficit of $1.5 million, the inaccessibility of the school's endowment funds for university purposes, and its chronic inability to meet admissions goals, "even with reduced standards of admissions," Kusch offered a comparative assessment: "The School is at a high level of mediocrity and lacks the distinction carried by MIT, Cal Tech, Stanford and a modest number of other institutions."[31]

Citing the negative impact on the university's already fragile morale, Kusch concluded that "it would be brash to attempt to remove the school from the Columbia University community." A month later, however, he alerted the president of plans afoot for a "possible consolidation of engineering schools in New York City," which, had it happened, would have relocated Columbia engineers to NYU's Bronx campus.[32]

McGill was sufficiently impressed with Kusch's administrative insights that in December 1970 he made him his university provost. The Nobel laureate lasted just six weeks in the job before submitting his resignation, effective July 1, 1971. He cited personal exhaustion as prompting the decision, though McGill later believed Kusch quit because he concluded the university was "headed for disaster, [and] wanted to get out." He left Columbia a few months later for a professorship at the University of Texas. In his farewell letter to SEAS Dean Hennessy, whom he liked, Kusch gave vent to his gallows humor and showed his familiarity with popular mysteries: "The SEAS is at something of a crossroads in a variety of ways. Isn't it at a crossroads, on a dark and thunderous night, that a stake is driven through the heart?"[33]

While Kusch credited Hennessy with understanding the financial plight of the university, unlike some other deans, and was impressed by his determination to trim his school's expenses, President McGill was less charitably disposed. Hennessy was, in the president's eyes, a captive of his department chairs, with neither the professional credentials nor fortitude to stand up to them. (McGill may have been privy to the crack the Columbia physicist Henry Foley made about Hennessy's administrative prowess: "Wes was never in a fight that he won.") Preoccupied in his first two years with seizing control of the finances of the university and, as he put it, "getting Columbia off the front page of *The New York Times*," required McGill to put the SEAS situation "on the backburner," though he identified engineering as a problem in need of his personal attention.[34]

HOW BAD WAS IT?

McGill began his presidency thinking that his administration had five years to put the university finances in order. Failure to do so, he believed and allowed privately, would be fatal. He also knew that he could not look to a capital campaign for help, with the one launched in 1966 aborted in the wake of 1968 halfway to its goal of $200 million and with most of the money raised coming from the Ford and Mellon Foundations. Further support from these foundations was unlikely while Columbia's viability remained in doubt. Talk of Columbia being folded into the then still growing SUNY system could be heard at the Faculty House.[35]

Among the first moves McGill made to limit expenditures was to impose a hiring freeze on faculty and administrators, while deans were directed to come up with ways for cutting staff through the consolidation of lines and the elimination of programs. Assistant professors were told their prospects for tenure were unlikely, prompting several to decamp at the first opportunity. In academic year 1972–1973, there were only thirty-seven formal tenure reviews conducted by the provost's office, less than half the number five years earlier. Not one involved an SEAS nominee.[36]

Another dramatic change in the university's operation involved the way in which tuition revenues were handled. Traditionally, these had gone to the individual schools, where they remained to cover school expenditures, with anything left over returned to the university (aka "Low Library"). With each school managing its own books in its own way, making accurate determinations of the university's overall financial situation was impossible. This all changed in 1972 when a new financial system was imposed on the schools. Henceforth all fees that had earlier gone to the schools would now go to the university, a portion of which would go back to the schools in the form of an annual allocation, with which the schools were expected to make do. Additional funding was to come from the schools' drawing down their restricted endowment accounts. This arrangement did not sit well with the strongest revenue-generating schools, the medical school and Lamont-Doherty. When McGill was later asked how the amounts allotted to each school were determined, he responded: "I decided them and then forced them down the throats of the parties."[37]

Unlike Lamont-Doherty Geological Observatory's director, Maurice Ewing, who resigned and left for the University of Texas rather than go along the new arrangements, SEAS's Dean Hennessy quietly accepted them. "Your allocation for 1973–74," Provost William Theodore de Bary wrote to Hennessy in 1973, "is $1,191,000. No increases in general income through 1978–1979 should be assumed."[38]

Full cooperation with McGill's "Salvation Plan" did not spare the engineering school from having its finances scrutinized by the president's auditors, who, in 1972, turned up a serious problem. The total cost of the construction of Seeley Mudd and Engineering Terrace buildings, some $15.3 million, had not been fully covered by the school's capital campaign. The shortfall was more than $3 million, for which the school remained responsible.[39]

In light of this discovery, Low Library gave over a floor of Mudd to the biology department, which in the closing days of the Kirk presidency had been promised its own building. Hennessy, McGill dutifully informed the SEAS faculty, "vigorously protested," citing understandings with the Krumb estate and the Mudd Foundation, but went along after securing assurances that the arrangement was temporary and the space would be returned to SEAS when the biologists got their own building. That building, it was further decided, would occupy the air space above Engineering

Terrace that Dunning had designated for his fantasized "Krumb Towers." The Sherman Fairchild Biology Building opened in 1976, but biologists stayed on in Mudd.[40]

Suddenly, SEAS seemed fair game for all manner of intrauniversity encroachments. In the fall of 1972, the chair of the mathematics department Edgar Lorch objected to SEAS faculty teaching differential equations in a popular evening course that dated back to the early 1960s and called upon Provost de Bary to halt the practice. The provost complied, confirming the math department's monopoly over the instruction of mathematics throughout the university. Meanwhile, the dean of Columbia College periodically called for either the merger of SEAS into the College or removal of SEAS students from the staff-strained College core curriculum. At this point *Pulse*, a normally even-tempered model of undergraduate journalism, editorially declared "Enough Is Enough":

> Ever since the engineering school has existed, the College has been dictating the policies of the University. Our School has always had to play second fiddle to College administration.... We urge the School's administration to show more courage to face such crises. Our School has been pushed around enough and it is now time for us to fight.[41]

It took three years before Low Library's budget officers provided the president with the numbers for a university-wide school-by-school financial analysis. Those for the engineering school were not good. They showed its "expenses over revenues" had been running for the past several years at around $3 million per year, with the costs per full-time student rising steadily. It was clear to anyone who saw these numbers that only by bringing about a rapid increase in enrollments and making sharp cuts in faculty could the situation be reversed. Dean Hennessy gave assurances to Provost de Bary that he was committed to both, but by then neither the provost nor the president believed him equal to the task.[42]

Finally, two more challenges facing Columbia at large in the early 1970s further complicated the lives of engineers and their embattled dean. The first came in November 1971 in the form of a letter from the Civil Rights division of the Department of Labor citing the university for failing to provide information attesting to its compliance with Title IX regulations concerning the employment of women. This followed on a complaint lodged by Columbia women employees citing statistical evidence they had compiled from catalogs and other university publications that made a prima facie case for longstanding gender discrimination. The university was now on notice that it would not only have to come up with numbers it had never felt the need to compile, but to undertake corrective action to address the underrepresentation of women in its employ that those numbers would, to a certainty, reveal.[43]

Underrepresentation of women among the university's faculties was beyond dispute. Of the university's thirteen professional schools, only the School of Nursing had enough women faculty to escape censure. The law and business schools had no women on their faculties in 1971. The engineering school had a faculty of ninety-five members, all male. Faced with the prospect that without remedial action the university risked the withdrawal of all federal funding, and taken aback by reports of egregious and longstanding instances of salary discrimination against women that came to light, President McGill assured the Department of Labor and women's groups on campus that corrective policies would be promptly put in place.[44]

This meant hiring more women faculty by Columbia's schools, even as most, including the engineering school, were under pressure to reduce faculty size. True, unlike several schools and departments in the humanities with wildly disparate gender ratios of students and faculty, the engineering school enrolled so few women that it was less open to the charge made of departments in the humanities, specifically English and art history, that they gladly accepted women as tuition-paying students, certified their degrees, and then refused to hire them as faculty. Unfortunately, SEAS, given the scarcity of female engineers, was also the least likely to come to the university's rescue with a flurry of women-faculty hires. It did manage to hire three women faculty between 1971 and 1973, all of whom later left before getting tenure.[45]

The other problem facing Columbia in the early 1970s was the deteriorating fiscal condition of New York City and increasing doubts about its governability and livability. Here, too, this problem affected the engineering school in a distinct way. Early in its history, the engineering school took pride in its location in the "center of American engineering." This claim was based on the presence of many of the largest engineering firms and of the corporate headquarters of many of the country's largest employers of engineers. A goodly portion of the school's graduates had traditionally secured employment at New York-based firms such as IBM, General Electric, and AT&T. But in the two decades after World War II the city lost half its manufacturing jobs as firms moved first to the suburbs of New York City and then dispersed further. The occupational attractions of the "Big Apple" had become decidedly less appealing, even as its quality of life for families with kids to educate became problematic. Even more than their colleagues in the arts and sciences, engineering faculty led the way in the exodus of Columbia professors from Morningside to the suburbs of New Jersey or Westchester. If already living in suburban Dobbs Ferry or Leonia, why not suburban Ann Arbor or Palo Alto or Berkeley? As for prospective engineering students or recruited faculty, why risk coming to a university in a city in such financial straits and psychological disrepair?[46]

"WE ARE MOVING"

In the fall of 1971 President McGill used the occasion of a problematic accreditation report from the Engineering Council for Professional Development on the SEAS nuclear engineering program to thrust himself into an ongoing review of the engineering school. He did so by having Provost de Bary direct Hennessy to prepare a statement on "the Aims and Goals of SEAS." True to form, the dean turned the job over to a committee of department chairs.[47]

The faculty committee, chaired by Ralph J. Schwarz, chair of Electrical Engineering, dutifully submitted its report in June 1972, only to have it summarily rejected. McGill later recalled de Bary being "outraged and astonished" by its vacuity. It was returned for a complete rewrite. The following spring Schwarz complied, with his *Academic Directions for SEAS—A Second Report of the Committee on Aims and Goals.* It was more self-critical than the first, acknowledging, for example, that only half the engineering faculty had external support for their research. It also admitted that Columbia's "very respectable rankings" in the four engineering fields covered in the 1964 ACE survey "had declined noticeably by 1969." However responsive to some of the criticisms of the first report, this second effort only reinforced the conviction within the university's central administration that the drastic action required if SEAS was to remain a part of the university would not come from within. One senior member of the engineering faculty, Professor of Chemical Engineering Elmer Gaden, said as much. While admiring of the committee chairman's efforts, Gaden voiced his belief that the *Report* "completely misses the mark" in failing to acknowledge the need "for a weeding out of our senior faculty." In the absence of the dean's active leadership, which Gaden said "evaporated" whenever Hennessy was challenged by department chairmen, Schwarz settled for a "least common denominator report" produced by "a committee to preserve the departments." Gaden left Columbia the following year.[48]

Kusch's prophesied "dark and thunderous night" seemed at hand. All through the spring of 1973, some SEAS faculty believed the closing of the school was under active presidential consideration. Others spoke of its "imminent demise." That NYU, for financial reasons, had just announced plans to merge its engineering faculty with that of Brooklyn Polytechnic Institute, effectively taking NYU out of engineering education, and that Yale, because of quality concerns, had recently dismantled its undergraduate engineering school, gave local credence to these apocalyptic views. On April 19, 1973, the SEAS undergraduate admissions committee reported a 14 percent drop in applications. Seven days later, the school's prospects seemed to touch rock bottom at the annual SEAS faculty meeting, the president in attendance and the provost presiding. (Faculty of a symbolic mind later recalled the meeting had to be moved to the law school because the elevators in Mudd had broken down.) Early in the meeting, Provost de Bary was asked by a faculty member if plans were afoot to close the engineering school. "No," he responded, "though your asking suggests maybe we should."[49]

What the president did instead was solicit outside evaluations of the school's standing and its future as set forth in the revised *Aims and Goals Report*. He did not hold back from the consultants his personal take on the school:

I have a feeling that our School of Engineering is standing still or per-haps regressing and the continuation of current policies will damage the School. We have no present intention of eliminating engineering at Columbia (as NYU did), but we seem to be drifting badly. I am trying to break out of this condition of benign neglect before the damage is done.[50]

McGill's consultants included Edward Cranch, dean of the Cornell engineering school, William O. Baker, head of Bell Labs, and Walter Rosenblith, provost of MIT. All three offered sharp criticisms of the school, but it was Gordon Brown, the just retired dean of the MIT School of Engi-neering, who, at Rosenblith's urging, responded to McGill's request with Kusch-like bluntness: "Your School of Engineering at Columbia has lost its way," he told Columbia's president. "No one at MIT takes the Columbia engineering school seriously. No one knows who the Dean is." What SEAS needed, Brown concluded, was "a Young Turk to shake up the School."[51]

The president called Hennessy to his office for his response. McGill later described his dean as "a genial fellow and hardworking," but at that meeting McGill characterized his position as "unusual," referring to his lack of pro-fessional credentials. The understandably defensive dean characterized his critics on the SEAS faculty as "malcontents who want my job so bad they can taste it." Fourteen months later, on September 6, 1974, McGill wrote to MIT provost Rosenblith: "I have retired the Dean, effective June 30, 1975. . . . Thanks for your help; we are moving."[52]

7

CATCHING A LIFT 1976–1980

Among my surprises at Columbia is the discovery of the degree to which our Engineering School is isolated from other parts of the University.

—DEAN PETER LIKINS TO THE SEAS FACULTY, OCTOBER 11, 1976[1]

The job was to build a great department, and unbeknownst to Columbia, to see what extent computing could change one of the great research universities. . . . I always had the picture that [computer science] was as important a discipline as physics or mathematics, but that certainly wasn't the vision of Columbia.

—JOSEPH TRAUB, ON HIS RETURN TO COLUMBIA IN 1979[2]

"I REALLY WAS AN ENGINEER"

MIT's "young Turk" dean, Gordon Brown, told President McGill his engineering school needed took some finding. A search committee chaired by Professor of Mechanical Engineering Ferdinand Freudenstein labored through Hennessy's last year as dean (1974–1975) and into the next, while Ralph J. Schwarz served as acting dean. Several factors impeded the committee's progress, not least the president's firm opposition to inside candidates. Credible outsiders could hardly view Columbia's shaky financial condition and SEAS's tenuous standing within the university without trepidation. Rumors put the number of candidates offered the position who turned it down between three and six.[3]

Fifteen months into the search the committee came up with name of Peter Likins, a professor of civil engineering and associate dean of the School of Engineering at UCLA. It helped that Likins had studied at Stanford (PhD engineering mechanics 1965) with Tom Kane, a CU PhD under Raymond Mindlin and an engineer of international standing, as did the fact that a onetime colleague of Likins's at UCLA, Steve Dubowsky, a Columbia PhD teaching at MIT, recommended him. It is also likely that McGill, who retained close personal ties throughout the California university system, made inquiries.[4]

In a 2012 interview, discussing his pre-Columbia career, Likins allowed, "I really was an engineer." The decision had been made at age ten when he happened on a biography of the chief engineer of the Panama Canal, George Washington Goethals. From Santa Cruz High School, Likins proceeded to Stanford, where he majored in civil engineering and mechanics, became a nationally ranked intercollegiate wrestler and married his high school sweetheart, Patricia. After graduation in 1957 he went to MIT to secure a master's degree and was there when the Russians launched Sputnik. He returned to California to work at the Jet Propulsion Lab for two years before returning to Stanford for doctoral studies. In 1964 he became an assistant professor in the UCLA School of Engineering, a joint member of the Astronautics Division and the Mechanics and Structures Division, even as he continued as a consultant for the aerospace industry. In 1972, now a full professor, he decided to try his hand at academic administration, accepting a one-year position as assistant dean of the engineering school. A year later he became associate dean. A year later President McGill came calling.[5]

Until McGill's visit in January 1976, a Likins appointment seemed a reach. He was thirty-nine and looked several years younger, had only two years of administrative experience, and was a lifelong Californian. The prospect of relocating his wife and six adopted children to New York was daunting. But apropos of a later career move, Likins said of his outlook, "When opportunity calls, I go." So when McGill offered him the Columbia deanship, he accepted. Six months later, Likins, his wife, kids, and dog, packed into their cross-country camper, reached New York in July at which point he commenced his duties as the tenth dean of the School of Engineering and Applied Science. His was to be the second shortest deanship up to that time, just four years, but also its most transformational.[6]

Despite having snagged themselves a young, energetic dean who "really was an engineer," the SEAS faculty in the summer of 1976 still had reason to lament their collective condition. A salary freeze imposed in 1971 had by 1974 left them the lowest paid engineering faculty in New York State. Lest the austerity imperative be lost on the faculty on the eve of the arrival of their new dean, McGill told them at its April 13 meeting that he had "informed the new Dean that the stress period within the University will continue and Engineering must raise money with its own hands."[7]

One bright spot on the gloomy fiscal horizon was that the undergraduate enrollment slump which had hit bottom in 1971 had by 1974 turned back upward. Graduate enrollments, too, which experienced a free fall in the immediate wake of 1968, had stabilized by 1973 and were now on the upswing. Low Library staffers monitoring the SEAS budget reported to the provost that "the figures seem to suggest that the slipping enrollment is reversed." The upward trend continued throughout the Likins deanship. Every year the school produced more tuition revenue than the year before, most of which went into a university-wide pool along with the tuition surpluses generated by the medical, law, and business schools, which was then used to prop up the deficit-running schools (principally the Graduate School of Arts and Sciences). Still, more each year came back to SEAS. Finding themselves among the university's income generators, if not quite a "cash cow," engineers enjoyed a novel taste of noblesse oblige as contributors to the president's "Salvation Plan," even as they strengthened their dean's bargaining position vis-à-vis the "tough guys" in Low Library.[8]

THE LIKINS DEANSHIP: "AT THE CUSP OF CHANGE"

"I am hard at work," the newly installed dean wrote to his boss Provost de Bary after seven weeks on campus, "to develop an understanding of Columbia's needs and resources." He then laid out his intentions, as if in explicit contrast with the university's traditional modus operandi:

In the spirit of unguarded candor that I want to maintain as a central characteristic of our relationship, I am determined to find a way to break out of this deepening rut. We cannot continue without serious deterioration of the quality of the faculty.... I intend to define for you a set of options that will enable us to begin to regain some measure of control of our destiny.[9]

By the fall, when he met for the first time with his faculty, Likins had identified a core problem that required his attention and their help: "Among my surprises at Columbia is the discovery of the degree to which our Engineering School is isolated from other parts of the University." This disquieting discovery was at least in part offset by the welcome he received from his faculty and also university administrators. It helped with the former that he had credentials as an engineer and an academic as well as energy and candor. McGill later acknowledged that he took to Likins not least because, unlike some other deans and senior administrators, he "could add a column of figures and get the right answer."[10]

FIGURE 7.1 Peter Likins (1936–), engineer and academic administrator, ninth dean of SEAS (1976–1980) and first engineer as Columbia provost (1980–1982); later president of Lehigh and University of Arizona. Photograph by Columbia *Spectator*.

Source: Columbia University Archives.

The new dean immediately overhauled the school's fundraising operation, successfully raising more money than his predecessor with less going to cover costs. Still, by the end of his deanship in 1980, annual alumni giving had only returned to levels commonly reached back before 1968. In the area of corporate giving, Likins got some help from President McGill, who, if initially dubious about his dean's prospects in tapping the business community, eased his way with a supportive call to the CEO of IBM. Likins came away with a gift of $100,000. By the end of the decade the school had struck up relationships throughout American industry that significantly increased the number of gifts, with many of these corporate–school relationships developed through a group he called Columbia Engineering Affiliates that endures to the present.[11]

FACULTY ENHANCEMENT WITHOUT GROWTH

What Likins did not do in the first years of his deanship was increase the size of the faculty. Additional resources that came the school's way went into increasing salaries of existing faculty. Here he bumped up against traditional compensation practices he deemed counterproductive to his goal of raising the competitive standing of the school. Some departments, mechanical engineering a case in point, operated on the principle that all salary increases should be across-the-board, everyone moving up by the same percentage. Likins called it "distributing the crumbs to the needy at the expense of the powerful." Yet his efforts to "elevate the salaries of our brightest stars" in the face of the "highly developed sense of community in some departments" were, he later acknowledged, only partially successful.[12]

Annual negotiations between the dean and provost over the size of the SEAS allocation, and even more over the apportioning of any tuition and gift income that exceeded projections, were often bare-knuckled affairs. In 1977, when some funds from the Percy K. Hudson estate earmarked for the engineering school were shifted to help pay for the East Campus dormitory complex, an exasperated Likins complained to McGill's chief budgeteer, Bruce Bassett: "How can I win this game if no one will tell me the score?" In 1978 Likins took on Low Library over who got to keep funds solicited from his newly organized Columbia Engineering Affiliates. This one he won.[13]

For all the squabbling, Dean Likins, then eighteen months on the job, responded to Provost de Bary's 1977 request for his assessment of the engineering school over the previous five years with the following: "By virtually every measure that might . . . reasonably be applied, the School of Engineering and Applied Science is on a rising curve showing growth and healthy progress over the past five years." Among the positive indicators he cited were the growth in enrollments even as the faculty grew smaller; increases in alumni giving and activity; research support "growing rapidly by all available measures." He then cited the school's newly formed plasma physics graduate program led by Professor Robert A. Gross as one of the strongest in the world. Likins's "Conspectus" closed with the summary statement: "Strategies for survival have given way to plans for building for the future."[14]

While the size of the faculty in the mid-1970s held at around 90, important additions occurred. The senior appointment of Mischa Schwartz in electrical engineering in 1973 followed on a shakeup at Brooklyn Polytechnic Institute where he had directed a nationally recognized program in telecommunications. Similar circumstances at CCNY that year led to the appointment of Donald Goldfarb as professor of industrial engineering and operations research. Both would play important roles in their departments and in the school, Schwartz as the founding director of the Telecommunications Research Center, and Goldfarb for his work developing powerful optimization algorithms and for two stints as acting dean. The few other senior appointments occurred in the well-funded School of Mines: Tuncel Yegulalp (1972) in mining engineering, Ponisseril Somasundaran (1974) in mineral engineering, and Nickolas Themelis (1980) in metallurgy. Christian Meyer, a specialist in concrete technology, joined the civil engineering department in 1979 as an associate professor. Meanwhile, junior hires cycled in and out, with only two, the mechanical engineer Richard Longman (1973), who went on to do pathbreaking research in space robotics, and civil engineer Gautam Dasgupta (1975), whose research is in engineering mechanics, securing tenure.[15]

Although it had taken eight years, three years longer than he projected at the outset of his presidency, President McGill informed the engineering faculty in the spring of 1978 that "for practical purposes the [university] budget is in balance." Both at that meeting and later he credited the professional schools with generating the revenues that made the turnaround possible. Then and later the schools' deans also said as much, with greater or less plausibility. But none had a better claim to having played an important part of the "Salvation Plan" than Columbia's engineers, and no dean greater reasons for saying so.[16]

With the university books in balance and the engineering school operating in the black, the faculty no-growth dictum was lifted and some long vacant positions were filled. Among faculty appointed at decade's end who were to figure importantly in the school's future was Gerald Navratil (1978) in applied physics, who would take a leadership role in the work of the Physics Plasma Lab. The real surge in hiring, however, occurred not in the standing departments so much as in a department just beginning.[17]

A SECOND CHANCE FOR COMPUTER SCIENCE

By moving SEAS into the novel position of being a school seen as helpful to the university, Likins put himself in a position to propose cooperative arrangements with Columbia schools that had not heretofore considered partnering with engineers. The way SEAS became home to computer science in 1979 is a case in point. The story of Columbia's squandering its early lead in computational science has been told. By the mid-1970s, MIT, Stanford, and Carnegie Mellon—"The Big Three"—each had departments of computer science, or in the case of MIT, a department of electrical engineering and computer science, with twenty or more faculty. Columbia's entire computer science activity in 1978 was limited to a tiny department of mathematical statistics, in the Graduate School of Arts and Sciences

(GSAS), which consisted of two senior and five junior faculty, and to the electrical engineering department, where two senior and three junior faculty included computing in their research interests. A proposed degree program in computer science lacked accreditation, while growing demand among students and faculty throughout the university for instruction in computing science went largely unmet.[18]

At Likins's urging, and with his sweetener of $1 million from the recently probated will of Percy Hudson's widow, plus another $600,000 from IBM, and with the enthusiastic cooperation of GSAS Dean George K. Fraenkel and Acting Provost Norman Mintz, an agreement was reached in the fall of 1978. The five faculty lines assigned to mathematical statistics would be transferred from GSAS to SEAS, where they would be joined with five lines in electrical engineering to form a free-standing department of computer science—in the engineering school. As expected, the plan was opposed by the electrical engineering department, which wanted to keep computing within its domain, while science faculty, who regarded computers as another "gadget," were fine with leaving them to the engineers. A pronouncement attributed to Rabi suggests Columbia's traditional skepticism toward technology had not vanished: "Any field that has 'science' in its name isn't." In any event, the decisions to make computer science a department of its own and to place it in the engineering school mark an important, if at the time unnoted, moment in the evolving history of the sciences at Columbia.[19]

With funding and faculty lines identified, the task then became finding the person to lead the effort to reclaim a place in the field of computer science. He—it being 1980 and computer science—would need to have scholarly standing in the field and administrative experience in a competitive academic setting. In Joseph F. Traub, the forty-seven-year-old director of the School of Computer Science at Carnegie Mellon University, the search committee quickly identified someone who met both criteria. Traub's 1959 PhD came from Columbia, where he had been a Watson Fellow in applied mathematics (1957–1959). Recalling his years at Columbia as a graduate student, he commented: "I do not feel that I got a good education. What I did get was unlimited access to computers." From Columbia he went to Bell

Laboratories, where he started the area of optimal iteration theory. Later, he and a Stanford PhD student, Michael A. Jenkins, devised a fast globally convergent iterative method now known as "The Jenkins–Traub algorithm" for polynomial zeros, which since being published in 1970 has become a standard in black-box polynomial root-finders. In 1971 he was appointed executive head of the Department of Computing Science at Carnegie Mellon University, which Traub later called "the epitome of a technology-centered university." When on sabbatical at Berkeley in 1978–1979, Peter Likins came calling.[20]

Discussions followed about Traub leaving Carnegie Mellon for Columbia, but it was at a visit to New York in March 1979 that specific commitments were made. These included a new building, start-up funds, and a target size for the envisioned department of thirteen lines. "Thirteen lines, and that's it," Likins said. Traub, who had grown up in the city and attended CCNY, was accompanied by his wife, the author Pamela McCorduck, upon

FIGURE 7.2 Joseph F. Traub (1932–), Columbia PhD in applied mathematics (1959); computer scientist, professor at Carnegie Mellon, and founding chair of SEAS Department of Computer Science (1979–1988).

Source: Columbia University Archives.

whom in a private meeting President McGill lavished his considerable Celtic charm. The deal was consummated.[21]

Traub arrived on campus in July 1979, where he was met by senior colleagues Jonathan Gross, Theodore Baskow, and Stephen Unger, plus six assistant professors drawn from the junior ranks of the now defunct department of math statistics and from the electrical engineering department. With the promised building still three years into the future, the newly constituted department of computer science opened for business in what had been part of a student cafeteria in Mudd. Everything has to start, or in this case, restart, someplace. This said, the timing was auspicious, not least because personal computers were just coming on the market and even technology-averse luddites could sense a seismic shift getting underway in the way we all go about our work and our lives.[22]

Another early instance of interschool cooperation initiated by Likins was between SEAS and the Business School. Here the collaboration was helped by the fact that the Business School dean, Boris Yavitz, was a graduate of SEAS (MS 1948). More crucial to the undertaking was securing the active cooperation of new members of the industrial engineering department with an interest in operations research to reach out to counterparts in the Business School, the result being a jointly administered university-wide program in operations research.[23]

A final instance of collaborative restructuring, this time primarily within SEAS but requiring the support of the physics department, was the creation of the department of applied physics and nuclear engineering (AP/NE). This required the cooperation of engineering faculty distributed among several departments with a common research interest in the emerging field of plasma physics, among whom Robert A. Gross was a prime mover, and a contingent of distinguished senior faculty whose careers centered on nuclear engineering, a field that Dunning had done much to promote at Columbia in the 1950s, but which was in national decline even before the Three Mile Island incident in 1979. These faculty included Herbert Goldstein, William Havens, and Leon Lidofsky. When Likins brought the controversial reorganizational proposal to the engineering faculty the expected grumbling occurred, only to be quieted by a senior colleague who would be adversely affected by the proposed reorganization, as recalled by a member present, as having said:

> We've had a dean who was smart but not honest;
> then we've had a dean who was honest but not smart;
> now we have a dean who is smart and honest:
>   Let's let him do what he wants.[24]

What made possible so many structural changes over the course of a four-year deanship was the relationship Likins had with his faculty. "Faculty meetings back then," Gerald Navratil recalled about the Likins era, "took the form of deliberative assemblies," at which the dean gave his reasons for some proposed action and then opened up the meeting for debate. Not every change he pressed for he got, but those he did had the force of transparency and consensus.[25]

The successful relaunching of computer science at Columbia, the founding of the applied physics department and the establishment of links between industrial engineering/operations research and the Business School all have had a beneficial effect on the engineering school long after Likins left in 1980 to become co-provost of the university. Within five years of its founding the SEAS computer science department had grown to twenty faculty and had secured departmental funding from the Defense Advanced Research Projects Agency (DARPA), putting it the exclusive company of MIT, Stanford, Carnegie Mellon, and University of California, Berkeley as a nationally recognized resource in computing science. The AP/NE department, which would undergo two subsequent name changes, along with computer science, electrical engineering, and the more recently established biomedical engineering department, are today the school's most highly ranked departments. The links forged between the Business School and the department of industrial engineering have been maintained and strengthened, especially in the hybrid fields of financial engineering and financial management.

Likins enjoyed immediate success connecting with students who were impressed with the seriousness with which he took the school's responsibility in providing effective instruction at both the undergraduate and graduate level. His youthfulness, energy, and openness to undergraduates allowed him to continue the student-friendly style of his predecessor Hennessy, with whom he shared the relevant qualification of being father to a large family, in Likins's case, a large and multiracial family, while further distancing the school from the corporate ways of Hennessy's predecessor, Dunning. During Likins's deanship the ethnic and gender diversity of the school increased significantly, with the percentage of women approaching 20 percent at the close of his deanship, double what it had been four years earlier. Though statistics on student ethnicity are lacking for the period, anecdotal evidence suggests that a similar doubling in the proportion of nonresidents and U.S. minorities occurred. Some of these increases are a consequence of larger demographic shifts, but surely some must be credited to what the dean described as his "deeply held convictions that equal opportunity is a moral imperative."[26]

For all the constructive changes made within SEAS during the Likins deanship, a more lasting consequence may have been that made on the university. "The single action on my part that had the most important impact on Columbia University was creating the Computer Science Department," Likins allowed in a 2012 interview, "but I think a more profound contribution in the long run was changing the perception of SEAS in the University." The perception that existed of the school on his arrival in 1976 had been fixed since the early years of the century: a small professional school, ideologically at odds with the prevailing academic ethos, but otherwise unobtrusive and in any case off on the margins of the Morningside campus. When Likins stepped down as dean in 1980, SEAS had been acknowledged by both Columbia's outgoing and incoming presidents as an integral and contributing part of the recovering university.[27]

Joe Traub made a similar point about the evolving symbiotic relationship between SEAS and the university in the years after his arrival. "The job was to build a great department," Traub said in a 1985 interview,

> and unbeknownst to Columbia, to see what extent computing could change one of the great research universities. . . . I always had the picture that [computer science] was as important a discipline as physics or mathematics, but that certainly wasn't the vision of Columbia.

These two visionaries, Likins and Traub, were equally convinced that Columbia's engineers were not only the object of change but a force in changing Columbia, with SEAS as lever.[28]

TRIGA REDUX AND RESOLVED

One of President McGill's complaints with Hennessy and the engineering faculty on his arrival in 1970 was that they had dumped the decade-long TRIGA nuclear reactor activation controversy on him to resolve. When the 1973 *Aims and Goals Report* made little mention of TRIGA, other than to call for its prompt activation, the school's avoidance strategy seemed confirmed. Hennessy had appointed a faculty committee, composed entirely of supporters of TRIGA and chaired by Professor of Applied Physics William Havens. Its 1974 *Report* predictably attested to TRIGA's safety and called upon the president to authorize its prompt activation. McGill's response to one TRIGA backer on the engineering faculty, Metallurgy Professor Daniel Beshers, cited offsetting concerns about the operating costs that would come with activation against the possible necessity of returning federal funds used to install TRIGA were the university not to activate. "I am concerned about the impact on the operating budget." In any case, activation would await the outcome of the neighborhood lawsuits pending in New York courts.[29]

McGill's concerns about TRIGA went beyond the budgetary. As he later acknowledged, he viewed the opposition to TRIGA as part of a broader cultural assault on science by those who saw little value in either scientific inquiry or the concept of academic freedom. In responding to faculty calling on him to reject activation, he declined in the absence of creditable evidence that TRIGA posed a health hazard and in the presence of engineering faculty who "would be prevented from engaging in what seems to be a legitimate instructional activity."[30]

There the issue stood in the spring of 1979, with one final antiactivation suit pending, which the university challenged, and McGill and the trustees giving the impression they were ready to authorize activation upon the suit's resolution. Then, on Wednesday, March 28, officials at the Three Mile Island nuclear facility in Pennsylvania reported an accident in the plant requiring an immediate shutdown. Operators could not assure state officials that there had been no nuclear contamination of surrounding areas. With the safety of the nation's other two hundred nuclear power plants in question, the activation of TRIGA in the midst of one of the nation's densest population centers had become politically toxic.[31]

On Sunday, April 1, the president informed the trustees of his decision against activation. The following day he met with the engineering faculty to communicate his decision. In advance of their meeting, the faculty passed two resolutions. The first, reflecting the overwhelming support activation retained among its members, declared that "capitulation to outside pressure would be unwise." The second, offered by Ralph Schwarz and more responsive to political realities, urged "a moratorium that would forestall either activation or dismantling the reactor." McGill later recalled the engineers' response to his decision to have the university withdraw from the last remaining law suit and proceed with the dismantling of TRIGA: "The engineering faculty did not forgive me for that. They did not say it was right, but they said they 'acquiesced.'"[32]

Later in that same interview, McGill drew a more bracing lesson from this "tough, very tough, civilized but tough, meeting over the TRIGA reactor." He contrasted the role of the engineering faculty in their battle over TRIGA with the battle over building a gym in Morningside Park eleven years earlier, when the faculty let President Kirk take the fall. This time had "the [engineering] faculty coalescing behind the president when he needed it." That one of the last crises Columbia faced in a decade when crises were the norm was resolved by the responsible actions of the engineering faculty in concert with the sorely put-upon president bode well for the "back from the brink" university and the revivified engineering school.[33]

In the spring of 1980 President-elect Michael I. Sovern appointed Peter Likins co-provost of the university. As dean of the law school, Sovern had worked with Likins on the Deans Budget Committee and valued his counsel. Only the perceived problem of having exprofessional school deans as both president and chief academic officer prevented Sovern from making the forty-three-year-old Likins his provost. Instead he divided the provost responsibilities among historian Fritz Stern, medical school dean Henrik H. Bendixson and later Robert Goldberger, and Likins. Even so, as Sovern acknowledged in his own case and Likins's, their appointments, coming from Columbia's professional schools rather than arts and sciences, represented "a departure from the university's prevailing culture."[34]

As it turned out, Likins stayed on as provost for only two years before accepting the presidency of Lehigh University. There, he later recalled, he encountered the reverse of what he had confronted at Columbia: a prevailing culture that privileged science and technology at the expense of the humanities and social sciences. After eighteen years at Lehigh he went on to an equally successful presidency of the University of Arizona (2001–2011). As for his years as SEAS dean, he said in a 2012 interview: "I had the good fortune of being at Columbia on the cusp of change. I left Engineering much more connected to other parts of the university than in the 1960s," he allowed, before insisting that "it is a mistake to say Pete Likins made this happen." Yet personal modesty should not obscure the virtually unanimous judgment of his colleagues three decades later that his deanship marked a historic turnaround in the school's fortunes. Within the university, among its peers and as a global resource, the school has since continued, albeit unevenly, on an upward path.[35]

8

UNEVEN ASCENT 1980–1994

[He assumes] leadership of a School that is very much on the rise at Columbia. I have confidence in his capacity to lead Columbia to the top of the Ivy League, where we belong.

—PRESIDENT MICHAEL I. SOVERN, ON APPOINTING ROBERT A. GROSS
THE ELEVENTH DEAN OF THE SCHOOL OF ENGINEERING
AND APPLIED SCIENCE, NOVEMBER 5, 1981[1]

Even as [the Dean] Auston airplane approached the cliff, the seeds of renewal were already discernible.

—CHRISTOPHER J. DURNING, 2012, ON THE SEAS SITUATION IN 1994[2]

B Y 1980, COLUMBIANS had reason to believe that the university had weathered the preceding fifteen years of "troubled times," when student protests, faculty discord, financial problems, and doubts about New York City's viability all pounded the 226-year-old institution. Cautious optimism about the future seemed an equally appropriate posture for its 116-year-old School of Engineering and Applied Science, which during the second half of the 1970s experienced a rare combination (last seen in the 1880s) of dynamic leadership at the decanal level, presidential approbation, and enhanced standing within the university. The departure on July 1, 1980, of President McGill to California marked the end of a crucial turnaround chapter in Columbia's history. His successor as Columbia's sixteenth president was Michael I. Sovern, who had been appointed university provost the previous year with the understanding that he would become the next president.[3]

Born in the Bronx, the son of immigrants, a graduate of Columbia College and Columbia Law School, and since 1958 a member of the Columbia law faculty (the youngest tenured professor at twenty-eight), and for ten years (1968–1978) its dean, Sovern was both the first member of a professional school faculty and the first Jew to become president of Columbia. While the second distinction received more immediate notice, the first may have been equally suggestive of changes afoot at Columbia.[4]

COMPUTER SCIENCE: BACK IN THE GAME

The first challenge for the newly constituted Department of Computer Science was meeting the pent-up demand across the university for courses in computer science. In September 1979 some two thousand undergraduate registrations resulted in classes of more than two hundred students. It soon became clear that the original commitment of thirteen lines would not begin to meet the demand. Nor would they satisfy the department chair's ambition to make Columbia computer science a presence beyond Columbia. The first tenure appointments were made in 1982 to two young foreign computer scientists, the Pole Henryk Wozniakowski and the Israeli Cornell-trained Zvi Galil. Both came on a semester-here, semester-there arrangement.[5]

Additional hiring opportunities developed when Traub and his three senior colleagues, Jonathan Gross from math statistics, and Theodore Bashkow and Stephen Unger from electrical engineering, informed the department's six assistant professors that they would be replaced by hires of the new department's choosing. In the end one assistant professor, Salvatore Stolfo, did survive what came to be called "the slaughter of the lambs." The replacement hires included John Kender from Carnegie Mellon, Michael Lebowitz from Yale, David Shaw from Stanford, and the Israeli UCLA-trained Yechiam Yemini; they were instantly dubbed "The Fabulous Four." Kathleen McKeown, a newly minted PhD from the University of Pennsylvania, was hired shortly thereafter. McKeown, Kender, and Yemini all subsequently secured tenure, while Shaw left academic life for the greener fields of finance.[6]

A hundred-thousand-dollar equipment grant from IBM in 1980 got the department up and going. In 1982, three years into the rebuilding process, came early validation that computer science at Columbia was back on the national map. The Defense Advanced Research Projects Agency (DARPA), then and now the largest federal underwriter of research in computer science, extended institutional support beyond Carnegie Mellon, Stanford, and MIT to include the University of California, Berkeley, on the West Coast and Columbia University on the East Coast. DARPA funds, which amounted over several years to $17 million before the agency stopped making institutional grants in 1985, allowed the department to make more junior appointments, including those of Gail Kaiser, Peter Allen, and Steven Feiner.[7]

In 1983 the Computer Science Building, tucked between Mudd and Sherman Fairchild, opened for business. A year later the department consisted of an instructional staff of nineteen members, with four thousand undergraduate enrollments, sixty PhD students, and $6 million in outside research support. In 1985 Traub was elected to the National Academy of Engineers, the first of what over the next two decades has grown to an impressive list of NAE-member Columbia computer scientists. That same year in an interview with the Institute of Electrical and Electronic Engineers, Traub offered the following assessment of how far Columbia had come in reasserting its standing in computer science:

> In my view, after the top three (MIT, Stanford, Carnegie Mellon), there is a group of universities that includes perhaps ten universities: Berkeley, Cornell, Yale, UCLA, Illinois, schools like that; that's not a complete list. We're in that group today.

Whether Columbia's computer department, in only its sixth year of business and still much smaller than any of the schools Traub listed, already belonged in this second tier was questionable; what was not in doubt was the heady company its chairman aspired to keep.

Meanwhile, Traub himself remained more active than ever in research, working on algebraic complexity and then starting the field of information-based complexity with is colleague, Henryk Wozniakowski. In addition, he started the *Journal of Complexity* in 1985 and served as the founding chair of the Computer Science and telecommunications Board of the National Research Council in 1986.[8]

The resources being committed to the computer science department inevitably generated resentment in less generously supported parts of the school. By 1985 computer science was, along with the Krumb-funded department of mining, metallurgy and mineral engineering, the school's largest. Moreover, computer scientists were, by their own estimate, a cocky bunch. The summary clearing out of inherited assistant professors had earned the department a reputation for Darwinian ruthlessness. Traub's style as chair has been described by admirers as that of a "benevolent dictator." Others allowed he could be demanding. A policy he brought from Carnegie Mellon of "pooling" the department's discretionary resources, where allowed, worked well enough when DARPA funding was being lavished on the department, providing start up support to incoming and unfunded junior faculty. But this spread-the-wealth policy became divisive when DARPA funds were interrupted after mid-decade, industrial support flowed in unevenly, and some department members self-identified as free-marketeers opposed the prevailing "socialism."[9]

In January 1986 Traub surprised his colleagues by announcing he was stepping down from the chairmanship and would be going to Princeton. Jonathan Gross, the only other full-time tenured member of the department, became chair until later that spring when Salvatore Stolfo, the sole survivor among the department's original junior faculty, secured tenure and was promptly conscripted into the thankless job of acting chair. To him fell the responsibility of wrangling with Dean Gross and university auditors in a joint effort to put the departmental books in order while trying to keep the peace. The pooling system was dismantled, with faculty now expected to secure and keep their own funding, along the lines of Harvard's famed school-by-school strategy of "every tub its own bottom," but here applied to individual bathers.[10]

After less than a year at Princeton, Traub decided to return to Columbia and so informed his departmental colleagues and university officials. Once back he was reelected department chair, but under new departmental rules that placed term limits on the position and under strained political circumstances. The department had acquired a reputation for being "a war zone." Traub's yearlong efforts as chair to return the department to the communal culture of earlier days, or, as an admirer and successor as chair characterized it, "to put the genie back in the bottle," failed. In the spring of 1988 he declined to stand for reelection, thus bringing to a close the department's remarkably successful if fractious founding decade.[11]

By then the department had several senior faculty from among whom to choose its next chair. The one chosen, Zvi Galil, was a surprise. He had come to the Columbia computer science department as a visitor on sabbatical leave from the University of Tel Aviv at the behest of Traub in the fall of 1982. He had received his BS and MS (his PhD was from Cornell) from the University of Tel Aviv, which his father had helped found in the 1950s, and where he had become a full professor in 1981 at the age of thirty-four. For several years after 1982 he alternated between Columbia and Tel Aviv, where his wife, a marine biologist, and son continued to live. While in residence at SEAS, he focused on his active research program and graduate teaching, otherwise taking little part in school affairs. In 1987 he was named to the Clarence Levi Chair in Mathematical Methods and Computer Science. Within the department he was politically allied with Traub but on good terms with those who were not. He had been chair of his Tel Aviv department and had some exposure to administrative labors through his professional affiliations, but his election to the chairmanship reflected less his administrative promise than the hopes of his colleagues that his personal skills could calm the roiling departmental waters.[12]

Galil proved a superb department chair, both administratively and in restoring comity to department relations, even as he maintained his research agenda. Upon completing a three-year term, he was unanimously reelected to a second two-year term, upon completion of which he resumed a program of full-time teaching and research. By then, Columbia's computer science department had still to break into the top fifteen in national rankings, a reflection of how difficult it had become for even the most ambitious department to move up in the rankings, especially a relatively small one making a late start in a rapidly expanding discipline. But it was pointed in the right direction.[13]

THE BRAVE NEW WORLD OF BAYH–DOLE

A legitimate criticism made of Columbia University throughout the middle half of the twentieth century, in fat years and lean, was that it was slow picking up on what the social commentator Michael Lewis called "the new new thing." And when it did get out in front—computing in the 1930s, and in the engineering school, nuclear engineering and bioengineering in the 1950s— Columbia all too often lacked the foresight and nimbleness to exploit its early entry. Thus the story of how Columbia University responded to the 1980 federal legislation known as the Bayh-Dole Act, which fundamentally changed the relationship between researchers, the university, and the government, offers a recent counter instance of Columbia quick off the mark and there at the payoff.[14]

The practice of Columbia faculty making their services available to governmental and industrial interests dates back to James Renwick in the 1820s. Of the three first faculty appointees to the School of Mines, Egleston and Vinton regularly worked for mining companies between terms, while Chandler held overlapping consultancies and retainers with all manner of corporate, industrial, and civic entities. Most of the other faculty of the school's first and second generation, among them the civil engineers Trowbridge and Burr and the electrical engineers Crocker and Pupin, regularly consulted with and carried out assignments for large engineering firms and electric power companies in New York City. Indeed, opportunities for such remunerated activities brought some engineers to Columbia in the first place. In the 1930s, to cite a few instances, Professor of Mechanical Engineering Charles Lucke worked three days a week at Babcock and Wilcox, his department colleague George Karelitz consulted regularly for Westinghouse, and Professor of Industrial Engineering Robert T. Livingston served as president of Long Island Lighting. In his 1950 history of the mechanical engineering department, Lucke praised "the close contact of the staff members with the engineering industry, primarily by part-time participation as consultants."[15]

Such contacts were not unique to Columbia's interwar engineers. MIT's famous "Tech Plan," for example, introduced in 1919 but informally operating since the 1890s, actively marketed the expertise of the MIT faculty as a resource available to industry. The difference was that MIT charged industrial giants such as Eastman Kodak and DuPont annual retainer agreements, with the income going to the institute; the arrangements between Columbia engineering faculty and their industrial employers were in the nature of private contracts, from which the university derived no income. This was also true of the work of Columbia's several testing laboratories, which operated under the aegis of a given engineering department and were staffed by engineering faculty and their students, the revenues used to keep the labs in operation. Even in the case of IBM's arrangements with Columbia, research space and equipment came to Columbia in the form of deeded gifts, and key faculty were put on the IBM payroll while the university itself went uncompensated.[16]

Faculty arrangements with outside companies and the income they derived thereby were never popular among the trustees, who periodically questioned their propriety, only to have them defended by successive engineering deans as necessary to attract and retain good faculty. [The practice may also have allowed the trustees to justify the lower salaries paid engineering faculty than those in arts and sciences.] Only in particularly egregious instances, where a faculty member's academic responsibilities went unmet or where the university found itself implicated in a legal entanglement, did a president or the board intervene to remind the faculty member that his primary obligation was to the university, not to what the university statutes reprovingly referred to as "unofficial employment."[17]

Faculty in the humanities with few opportunities for lucrative term-time employment complained about the resultant inequity, while some "pure" scientists saw such "downtown" employment as a lesser form of prostitution. Again, it seemed not to matter that such employment was at least as commonplace among engineering faculty at MIT or Stanford, where it implied no such moral onus; at Columbia it carried the mark of the devil.[18]

A similar hold-your-nose-laissez-faireism characterized Columbia's institutional position on income derived from patents held by faculty inventors on the university's payroll. A patent filed by Columbia engineer Michael I. Pupin for his 1891 invention of the inductance coil, subsequently used in the transmission of voice signals in long-distance communications, was sold by the inventor to AT&T in 1901 for a percentage of the profit AT&T earned from the future sale of "Pupin coils." Over the next three decades this arrangement made Pupin a millionaire, while Columbia University, his primary employer and owner of the laboratory where the coil was invented, made no claims to any part of his millions.[19]

A similar instance was of a 1934 patent filed by Edwin H. Armstrong, a 1914 graduate of the Columbia engineering school, for his invention of low-noise FM radio, when he was an unpaid honorary professor of electrical engineering and conducting his research in the Marcellus Dodge Laboratory in Pupin Hall. Again, Columbia made no claim to the income he derived from his patent, which involved a protracted legal challenge mounted by attorneys for the Radio Corporation of America (RCA) and its head, David Sarnoff. In this case, however, the professor of physics and future Nobel laureate Charles Townes did address what he regarded as Armstrong's ethical shortcomings in keeping the money from his invention instead of, as Townes did with the patent rights to his invention of the maser, conveying them to the nonprofit Research Corporation for the underwriting of basic research. Either way, Columbia came away empty-handed.[20]

There the matter stood until the spring of 1980 when Congress passed the Bayh-Dole Act. This legislation followed extensive discussions between research university officials and legislators anxious to help universities finance scientific research but without the federal government underwriting an ever greater share of the cost. The idea was to have federal funding agencies transfer whatever claims they had to income from patents to the universities where the inventors worked and the inventions occurred. This gave the university an immediate incentive to secure, along with the inventor, patent licenses to turn the invention into an income stream benefiting the inventor, the school, and the university. The original distribution arrangement approved by Columbia trustees worked on a sliding scale, whereby the faculty share declined as the total income grew.[21]

Columbia, led by its lawyer president Sovern and medical researcher provost Goldberger, wasted no time doing just that. The first and still the most lucrative instance of patent licensing by Columbia involved Professor of Biochemistry and Molecular Biology Richard Axel on the medical school faculty, who invented and patented a procedure for the co-transformation of genes, which permitted stable introduction of any gene into cultured cells. During the twenty-one years of the patent's life it made Dr. Axel a wealthy faculty member, the Medical School the recipient of millions of dollars in research funding as its share of the Axel bonanza, and earned Columbia University hundreds of millions in patent revenue. In keeping with the entrepreneurial ethos that inspired the legislation, Columbia trustees, with the urging of President Sovern and Goldberger's successor, Provost Jonathan Cole, approved a plan by which that income would be used across the university to stimulate still more research that had "commercial potential."[22]

Shortly after Axel was perfecting his DNA-splicing procedures uptown, over in the engineering school pioneering work was under way in the field of video compression by Professor of Electrical Engineering Dimitris Anastassiou. The patent stemming from this invention, which took commercial form as a critical part of the MPEG-2 standard used in DVD players and recorders globally, allowed Columbia to become part of MPEG LA, the patent pooling organization that developed many other technical standards over the ensuing years. Income from this arrangement over the twenty-one years of patent protection supported Anastassiou's research, provided his department and the engineering school with a significant income stream, and added many more millions to the annual total of university income from patents.[23]

Beginning in the mid-1980s and thereafter, Columbia has been among the top universities every year in the amount of patent and licensing income as a share of total income. And as Columbia's engineers increasingly direct their research into fields as diverse as medical devices, diagnostic tests, clean energy, and imaging, they are likely to become responsible for generating an increasing share of this income. Perhaps second only to the medical school in its likelihood to produce research "with commercial potential," the engineering school takes on an importance in the brave new world of Bayh-Dole not lost on any but the most unreconstructed technology-averse Columbian. Without adopting Gordon Gekko's anthem, "Greed is good," we have come a long way from judging the value of faculty research by I. I. Rabi's test: "Does it bring you nearer to God?"[24]

THE GROSS DEANSHIP

Robert A. Gross, "Bob" to all, had come to Columbia in 1960 as a tenured member of the mechanical engineering department, following undergraduate studies at Penn, a PhD in applied physics from Harvard in 1953, and six years at Fairchild Engine and Airplane Corp., where he became an expert in supersonic combustion. Two years later, along with his colleague the applied mathematician C. K. ["John"] Chu, he cofounded the Columbia Plasma Physics Lab. Following a schism within mechanical engineering in the early 1970s, he and Chu left the department to become professors in the program of engineering science. Upon the founding of the Department of Applied Science and Nuclear Engineering in 1978, he became its first chairman. His identification as the school's leading "applied scientist," as opposed to being one of its "real engineers," cost him inside support for the deanship back in 1975–1976. But with his standing as one of the country's experts in fusion technology, his tested administrative skills, the tempering of his earlier us-against-them style, and the strong endorsement of Likins, Gross's selection was well received by most SEAS faculty.[25]

On November 5, 1981, in announcing the appointment of Gross as the eleventh dean of SEAS, President Sovern wrote to the acting dean, Ralph J. Schwarz, that Gross was expected to provide "leadership of a School that is very much on the rise at Columbia." He then gave the new dean his marching orders: "I have confidence in his capacity to lead Columbia to the top of the Ivy League, where we belong."[26]

BIOENGINEERING REDUX

While supportive of the continued growth of computer science, Gross directed his early programmatic initiatives elsewhere. As with computing, Columbia had an early history in bioengineering that failed to develop beyond the "lone wolf" efforts of individual faculty. The first work at

Columbia in this as yet unacknowledged field was done by Michael I. Pupin, who in 1906 helped develop the X-ray tube, which revolutionized the practice of internal medicine by allowing selective scrutiny of the inner human body without surgery. A half century later, three members of the engineering faculty—Richard Skalak in civil engineering and Elmer Gaden and Edward Leonard in chemical engineering—pursued research agendas that applied engineering principles to the human body. Gaden's 1954 Columbia dissertation involved a process of penicillin production that could be accelerated by increasing oxygen during production; Skalak, a Navy V-12 graduate of Columbia who returned after the war, began in the 1960s to apply his engineer's experience with liquid flow patterns to blood movement in human bodies; Leonard's early work focused on improving the functionality and safety of artificial kidneys.[27]

In the 1960s Skalak teamed up with Shu Chien, professor of physiology and biophysics at the College of Physicians and Surgeons (P & S). They worked together at Columbia until reaching retirement age in 1987, when both went on to the University of California, San Diego, where they continued their fruitful collaborations on the biological implications and responses to implants. But their partnership-of-equals was the exception. More often the engineers were regarded by their MD counterparts as technicians, necessary to the funded research but hardly coprincipals. In 1967 President Kirk and Medical School Dean Merritt announced the creation of the Columbia Bioengineering Institute, a joint venture between the medical school and SEAS. Five members of the SEAS faculty were among the institute's founding members, but Professor and Chair of the Department of Physiology Dr. William L. Nastuk was designated chairman. Leonard served as the institute's secretary. Skalak expressed dissatisfaction with the power arrangements from the outset, while Leonard's minutes of the early meetings suggest that the major business of the bioengineering institute was "how to balance the status of engineers and physiologists." Four years after its launch, President McGill dismissed the institute as "a patchwork of interesting, narrowly defined enclaves, serving no broad need of our faculty and our student body, and no overriding public interest."[28]

Efforts to establish bioengineering at Columbia on a firmer foundation were renewed in the early 1980s, this initiative coming from the medical school's dean, Dr. Donald Tapley. Discussions between Tapley and Gross produced an agreement that the engineering and medical school should make a joint senior appointment in bioengineering to lead such an undertaking, with funding from the medical school. A search committee quickly settled on Van C. Mow, then the Clark Crossan Professor of Engineering at Rensselaer Polytechnic Institute. The son of a general in Chinese air force of Chang Kai-Shek's Nationalist Chinese government, Mow had come to the United States as a young boy shortly after World War II, and along with his two brothers, attended RPI, where he earned a BS in aeronautical engineering (1962) and his PhD in applied mechanics (1966). After a postdoctoral fellowship at the Courant Institute of Mathematical Sciences at NYU, he worked briefly for Bell Labs and then returned to RPI as a junior member of its department of mechanical engineering. His research was increasingly directed toward applying engineering principles to the human skeleton, which brought him into fruitful contact with medical researchers and orthopedic surgeons throughout the Albany area and at the Harvard Medical School. He also developed a collaborative relationship with an RPI engineering colleague, Professor of Mechanical Engineering W. Michael Lai.[29]

Mow's courtship with the Columbia medical school began in 1981 at meetings of the Orthopedic Research Society, of which Mow was the first PhD president, with an offer from P & S's orthopedics department chair, Frank Stinchfield. Stinchfield's opening ploy: "We'll pay for everything." Mow declined because Stinchfield could not assure him access to graduate students from the engineering school. There matters stood until 1985 when Columbia resumed discussions, but this time about a joint appointment as professor of orthopedic surgery at the medical school and professor of mechanical engineering in SEAS. For Mow, who had in the intervening years further solidified his reputation as a leading figure in the emergent field of biomechanics, accepting meant exchanging the certainties of the engineering professorship at the technology-centered RPI for the uncertainties inherent in a joint appointment at two schools not known for a history of effective collaboration, and at a university with a technology-averse past. But accept he did, because of the standing of Columbia's medical school and the ensured access to graduate students that the SEAS appointment provided.[30]

The medical school demonstrated its commitment to future collaboration with SEAS by proposing an appointment of Michael Lai that was identical to Mow's. The remaining sticking point turned out to be on the SEAS end. When Dean Gross informed the chair of the mechanical engineering department, Herbert Deresiewicz, of the pending deal, he was told that he and the other two senior members of the department (Ferdinand Freudenstein and Rene Chevray) were opposed to the appointments, not wishing to have their department become home to the "soft" kind of engineering they identified with bioengineering. Only Gross's insistence, and the fact that the lines were being paid for by the medical school, secured the appointments. Still, the coolness with which they were received persuaded Mow to focus his early energies on his laboratory up at P & S, while coming down to SEAS for his classes. Lai made a more concerted effort to make himself a home among his departmental colleagues in mechanical engineering, eventually serving as department chair. But with Mow and Lai now at Columbia, and Edward Leonard already in place, the prospects for a more institutionalized arrangement for biomedical engineering brightened.[31]

Another programmatic initiative of the Gross deanship involved the electrical engineering department and two of its senior members, Mischa Schwartz and Thomas E. Stern. Stern, an MIT undergraduate and PhD in electrical engineering (1956), had come to Columbia in the early 1960s and was department chair in 1973 when he recruited Schwartz. A graduate of Cooper Union (1947) with a PhD from Harvard (1951) in applied physics, Schwartz spent twenty years on the faculty of Brooklyn Polytechnic Institute, where he headed up BPI's nationally recognized telecommunications group. Financial difficulties facing BPI in the early 1970s influenced his decision to come to Columbia, where his cousin Melvin Schwartz was a professor of physics and later Nobel laureate.[32]

In 1984, when the National Science Foundation departed from its usual "only-basic-science" ways and announced a competition for establishing engineering research centers, and at the urging of Dean Gross, Schwartz and Stern concluded that "we had no choice but to write a proposal." Theirs, one of forty-two submitted to NSF, was one of six proposals funded the first year and the only one in telecommunications; it focused on the integrated transport of broadband information in voice, data, image, and full-motion video formats. The initial five-year grant was for $3 million. Schwartz served for three years as the founding director of the Center for Telecommunication Research (CTR). The grant was renewed in 1988, this time for $5 million, and, in 1991, for an almost unprecedented third time, for another $3 million. During the center's eleven years of operations (1985–1996), it underwrote a substantial portion of the salaries of between twenty-five and thirty faculty, provided research support to eighty-five graduate students, and effected both instructional and research collaborations among faculty from several engineering departments. These collaborative arrangements were later extended to include the Schools of Business, Journalism, and Law.[33]

Another early instance of cross-faculty collaboration that has since expanded was the Joint Services Electronics Program in quantum electronics operated by the Columbia Radiation Laboratory, to which engineering faculty first became participating members in the 1970s. When the program closed in the late 1980s, two of its leading members, Richard Osgood of the electrical engineering, and applied physics and applied mathematics departments and George Flynn of the chemistry department, continued the collaborative effort. The Radiation Lab secured an NSF grant to organize a Materials Research Science and Engineering Center, with Irving P. Herman of applied physics and applied mathematics its founding director; an IBM-mediated Center on Microelectronic Sciences, with Osgood as director; and a Department of Energy–sponsored Molecular Sciences Institute, with Flynn as director. In the 1990s the Radiation Laboratory was renamed the Center for Integrated Science and Engineering (CISE), with Flynn its first director, and has since hosted other cross-faculty centers, including an NSF Nanoscale Science and Engineering Center, a Department of Defense–funded center for Graphene Materials and Devices, and a Department of Energy–funded Energy Frontier Research Center. Herein rests the future.[34]

SCHAPIRO CENTER: EXPENSIVE BOONDOGGLE OR PRICELESS ACCOMMODATIONS?

The Center for Telecommunications Research was also slated to become the first tenant of the Morris A. Schapiro Center for Engineering and Physical Science Research (CEPSR), the building of which remains a legacy of Robert Gross's tenure as dean. First and foremost among his priorities was finding more space for the engineering school. Without it, he reasoned, SEAS could not expand its faculty. Without more space SEAS could not compete for grants that required commitments of thousands of square feet of modern laboratory space. In 1984 Gross estimated that SEAS had a backlog of between $35 and $50 million in grants contingent on finding space to undertake the research. Meanwhile, federal agencies, once generous with their support for buildings, had stopped accepting proposals that involved new construction.[35]

In 1985 opportunity came knocking. Governor Mario Cuomo announced a plan by which New York State, through its Dormitory Authority, would make available loans to private educational institutions prepared to use these funds to construct academic facilities, not just dormitories. President Sovern and Dean Gross proceeded to Albany and came away with assurances of a $40 million interest-free loan mortgage, payable in thirty years, to build an engineering research center. Expanding on Likins's success attracting corporate support and alumni gifts, Gross was confident the additional cost of the building could be covered by corporate and alumni donors to an SEAS building fund, including the pledged gift of $10 million from Morris A. Schapiro (CC 1924, BS engineering 1926), when combined with projected direct costs from funded research carried on in the building and expected help from the university. Plans were drawn up for a twelve-story building to be located on the northern edge of the campus between Mudd and Pupin Halls; groundbreaking was scheduled for January 1988.[36]

Three months before the groundbreaking shovels were distributed, the stock market experienced its worst collapse since 1929. Between October 14 and October 19, 1987, the Dow Jones Industrial Average plummeted

508 points, losing 22.6 percent of its total value. While the overall economy recovered quickly, the negative impact on corporate and alumni giving to Columbia persisted into the 1990s, casting a financial pall over the last years of the heretofore buoyant Sovern presidency. Its more immediate impact, however, was felt at SEAS, where a substantial shortfall in corporate and alumni giving left it short of revenues, even as it was saddled by the

FIGURE 8.1 Schapiro Hall—Center for Engineering Physical Science Research (CEPSR)—opened in 1991. Built at the urging of SEAS Dean Robert A. Gross (1980–1989); underwritten in part by Morris A. Schapiro (CC 1922, engineering MS 1925).

Source: Columbia University Archives.

university with annual mortgage payments on Schapiro Hall of $2 million for as far as the eye could see. Recriminations about responsibility for this reversal of fortunes were directed at Gross and led to his resignation.[37]

The financial problems with Schapiro, along with the inevitable grousings about aspects of the building's design (what recent Columbia building has avoided them?), made for a downbeat ending to Gross's nine-year term as dean. Yet there was much to commend it, both programmatically and in the provision of student services. Even what was judged at the time by some to have been a rash move to commit the engineering school to a building project without the resources in hand to pull it off, in retrospect can be seen as a courageous decision to expand and innovate. Rather than the consequence of an "edifice complex," the audacious building of Schapiro reflected not only Gross's determination but that of faculty and students, and in keeping with the mission of his predecessor, it served to move Columbia's engineers from the periphery to the center of university life. Financial exigencies brought on by unlucky timing may have momentarily stalled that transit, but everything about Bob Gross's deanship was directed to that end. The Schapiro Center for Engineering and Physical Science Research (CEPSR), ex-Dean Galil declared two decades after its opening, "was absolutely crucial to any of the growth the School has experienced since."[38]

WOMEN, FINALLY IN NUMBERS

When Columbia came under federal scrutiny in the early 1970s for its employment practices with respect to women, the university's story was mixed. Its graduate programs and professional schools had long been open to women students, with only the undergraduate College retaining a male-only policy. This exception was mandated by the affiliation agreement dating back to 1900 with Barnard College, which is, as the writing on its Broadway gate proclaims, the "undergraduate college for women of Columbia University."[39]

As noted earlier, SEAS enrolled its first woman student, Gloria Reinish (née Brooks), who entered with advanced standing in February 1943 and graduated in 1945. Of the next four women to enroll, three of them transfers from Barnard, only Anna Kazanjian stayed on to graduate with a BS in 1949 and an MS in 1952. She married her classmate, Guy Longobardo (BS 1950, MS 1952), before embarking on a distinguished career in aerospace engineering, the design of submarine navigation systems, and as an executive of Unisys Corporation. She also served on the Columbia Board of Trustees.[40]

No entering undergraduate engineering class between 1946 and 1966 included more than a half-dozen women, with women graduates any of those years counted on a single hand. Women represented an even smaller percentage of graduate students, most of whom were underwritten by corporate employers who had yet to show much inclination to hire women. Between 1953 and 1968, SEAS awarded a total of 1,410 undergraduate degrees, only 30 of which went to women.[41]

The principal complaint lodged against Columbia by feminists and critics was less its record enrolling women students than hiring women faculty. Indeed, one of the most damning statistics gathered by the Columbia Women's Liberation (CWL) and published in its 1969 *Report of Committee on Discrimination Against Women Faculty* was the disparity between the percentage of women receiving advanced degrees from Columbia departments and that of women faculty in the same departments. Several departments in arts and sciences in the late 1960s, among them French, psychology, anthropology, and philosophy, awarded between a third and half their PhDs to women, while these departments had no women faculty. The first woman to teach at SEAS was Elna Robbins (née Loscher), who had graduated in 1951 with a BS in industrial engineering, having been steered into that field by Associate Dean Hennessy. "Industrial engineering is about people," Robbins remembers him advising her and other women students. As a graduate student in industrial engineering in 1955 she was hired to teach a section of the department's introductory course. It was another sixteen years before a woman was appointed to a regular faculty position. That distinction came to Carolyn Preece, a Cambridge-trained metallurgist appointed to an assistant professorship in the Krumb School of Mines in 1971. She was followed the next year by Barbara Quinn, hired as an assistant professor in mathematical methods and operations research. Preece remained only a year before moving to Stonybrook, where her husband taught, and Quinn stayed for three years. Maryann Farrell (later Farrell-Epstein), who had been a graduate student in chemical engineering, was hired upon completion of her PhD when she became an assistant professor in chemical engineering in 1972. She stayed until 1981, when she too left without tenure.[42]

The appointment of Kathleen R. McKeown in 1983 as an assistant professor in the four-year-old department of computer science thus marks the close of a decade of false starts as regards SEAS's hiring of women faculty. McKeown, a Brown undergraduate majoring in literature and only turning to computing late in her junior year, had gone on to the University of Pennsylvania to write her dissertation on natural-language computing. In 1989 she became the first tenured woman on the SEAS faculty. Still, early informal leave arrangements made with successive department chairs to accommodate child bearing and family responsibilities had long-term negative consequences for her salary, as compared with her male contemporaries. Two years behind McKeown, Gail Kaiser, trained at MIT and Carnegie Mellon and with research focused on software systems, was appointed assistant professor in computer science in 1984. She received tenure in 1991, when she proceeded to have a child. When asked why both these appointments occurred in the same department, the chair of computer science at their hiring, Joe Traub, allowed: "It's nice to have very smart women here."[43]

By the early 1990s the department of metallurgy, mineral engineering and mining had two women faculty members. The first was the distinguished senior metallurgical chemist Gertrude F. Neumark, a Barnard College AB (1949) and Columbia PhD in chemistry, and for many years a principal researcher at Phillips Electronics. Appointed in 1985, she retired from SEAS in 2000 as the holder of an endowed professorship. The second was the metallurgist Siu-Wai Chan, whose research is in polycrystalline materials; she was appointed in 1992 and received tenure in 2000. Meanwhile, three other women were appointed to assistant professorships in the 1990s but left without securing tenure, another indicator at century's end of the engineering school's approaching gender parity.[44]

REIMAGINING THE AUSTON DEANSHIP

Two decades after its close, the record of the brief deanship of David H. Auston (1991–1994) remains mixed among those who experienced it. Auston himself has allowed that if judged by his popularity as dean, "I would not be considered a success." By two other measures, however, the capacity to identify the school's structural problems and the willingness to propose radical solutions to them, his deanship deserves reconsideration.[45]

A Canadian by birth with degrees from the University of Toronto and the University of California, Berkeley (PhD 1967), David Auston had been at Columbia only three years when he became dean. Prior to accepting appointment as professor of electrical engineering and applied physics in 1987, he had spent two decades at AT&T's Bell Laboratories in Murray Hill, New Jersey, in several research and administrative capacities. He was elected chair of the department of electrical engineering in 1989, the same year he was elected to the National Academy of Engineers, the sole member of his twenty-one-person department and only one of a handful of other SEAS faculty at the time to be so recognized. Despite the brevity of his time at Columbia and absent any extended experience in academic administration, Auston was nominated with the approval of President Sovern and on January 1, 1991, he assumed the office and responsibilities of the twelfth dean of SEAS.[46]

Part of what distinguishes one deanship from another is timing; Auston's was not opportune. In 1991 the university found itself with a short-lived but nonetheless painful budgetary shortfall. The ensuing efforts by the central administration at across-the-board belt-tightening led to a protest from department chairs in the arts and sciences. Meanwhile, President Sovern was preoccupied assisting in his wife Joan's losing battle with cancer.[47]

Columbia's financial strains were particularly acute at SEAS, the product of three interrelated factors. One was the need to pay for the as yet unnamed building under construction. Two income streams had been earlier identified as sufficient to cover its annual carrying costs—an increase in private gifts including a naming one, and increased recoveries from indirect costs charged on federal contracts—but neither had done so as of 1991. This left the school covering the carrying cost with operating funds. Auston inherited a school budget with a $1.6 million deficit.[48]

The second factor was a mismatch among the school's presumptions to disciplinary comprehensiveness, departmental autonomy and its modest size. Of its eight departments, only three—applied physics, computer science, and electrical engineering—had faculty numbers in double digits, while four others—chemical engineering, civil engineering, industrial engineering, and mechanical engineering—were perilously close to falling below what accrediting agencies considered an acceptable "critical mass." As for the Krumb School of Mines (aka department of mining, metallurgy and mineral engineering), its very existence at an urban university seemed justified only by a reading of the terms of the Krumb gifts which precluded underwriting other parts of SEAS.[49]

It was the decided view of the new dean, at the urging of the president and provost, that departmental consolidation and the attendant cost savings should be the order of the day. One year into his deanship, Auston sent a letter to all members of the SEAS faculty outlining his plan for departmental reorganization. All departments were told of the possibility that they would be merged with or subsumed by another department. The message conveyed to the six-member chemical engineering department was unequivocal: it was to be eliminated. The dean also informed the mining, metallurgy and mineral engineering department (aka the Krumb School of Mines) that its monopoly over the use of the Krumb endowment had ended.[50]

On notice, the chemical engineering faculty, led by Chairman Jordan Spencer, promptly launched a campaign to arouse alumni sentiment against the closure. Letters went forth from prominent graduates to the dean, and meetings were held to decry the proposed action. Students majoring in chemical engineering were also heard from; the *Columbia Spectator* weighed in with an editorial opposing the plan. For their part, members of the Krumb School faculty questioned the legality of the dean's plan to "raid" the Krumb endowment. Faculty in the civil, industrial, and mechanical engineering departments experienced a collective sympathetic shudder. "There but for the grace of God . . ."[51]

Auston had good reasons for targeting these two departments. Chemical engineering had limped along since the mid-1970s in the absence of effective leadership. It had trouble recruiting junior faculty and more trouble getting tenure for those it did recruit, thereby acquiring a reputation of "chewing up junior faculty." In 1992 the department had six members, down from a dozen a decade earlier. Student enrollments experienced a comparable decline. Its reputation within SEAS and among its junior members was of a "dysfunctional" department. Meanwhile, the seven-member mining, metallurgy and mineral engineering department continued to exist under the cloud of being the richest department in the school with the fewest students. In a good year the department graduated four or five undergraduate majors.[52]

With the chemical engineers protesting and Krumb faculty threatening to blow the whistle on the "raiding" of the Krumb endowment, Auston came up with a Rube Goldbergian variation on his original plans, but still directed to cost-saving departmental consolidation and freeing up the Krumb endowment. He proposed merging the chemical engineering and mining, metallurgy and mineral engineering department into a "department of materials science, chemical engineering, mining." The two departments reluctantly agreed, and because the Krumb faculty outnumbered the chemical engineers seven to six, the department chairmanship went to one of theirs, the La von Duddleson Krumb Professor of Mineral Engineering, Ponisseril Somasundaran. As chair, Somasundaran insisted on keeping two sets of departmental books, one for expenditures on mining, metallurgy and mineral engineering activities he deemed properly chargeable to the Krumb endowment, and one for expenditures on chemical engineering, which he deemed not. Hard to say who this action angered more, his chemical-engineer colleagues or the dean. There the matter unhappily stood throughout Auston's deanship.[53]

The dig-in-their-heels response of the chemical engineers and miners was a variation of that taken by the rest of the SEAS faculty: all assumed a defensive crouch and many refused to countenance, much less cooperate with, any of the dean's other consolidation plans for the school. The continuing financial strains under which the school operated, which required a one-year freeze on faculty and staff salaries and a stall in hiring, did not make matters easier. Meanwhile, Auston brought the school's budget into balance and secured a naming gift for the newly opened building from Morris Schapiro (ME 1924), which thereafter reduced its annual carrying costs. These achievements won him the respect of the outgoing president Michael Sovern and the incoming president George Rupp, both of whom put Auston on presidential commissions, a strategic planning exercise, and a commission on reordering the university's finances, which the engineering dean chaired. Auston looked upon these university-wide venues as opportunities to counter what he sensed was "the low esteem in which the engineering school was held." Much like the other outsider, Peter Likins, upon his arrival on campus back in 1976, Auston came to the deanship believing the school to be "isolated and insular," and, as with Likins, part of his job was to change that perception and the reality.[54]

Such views did little to assuage his immediate constituency. Two years into his deanship the SEAS faculty circulated a resolution of no confidence and communicated it to him. By then some faculty had the impression that the dean "didn't care about the place," others that he viewed SEAS "as a stepping stone." In May 1994, officials at Rice University announced that Auston had been chosen to be their provost, effective August 15. Several SEAS faculty have since recalled their reaction to the news as one of relief.[55]

In retrospect, Auston's consolidation plan was maladroitly pursued and effectively undercut by faculty opposition. It cannot be said to have identified nonexistent problems. Both chemical engineering and the Krumb School departments underwent more radical changes under Auston's successor, but, as will be described in the next chapter, with less rancor and more cooperation. The claims of Auston's deanship to a favorable reconsideration rest on more than a plea of "good idea/bad execution" in the instance of departmental consolidation; they rest also on two other initiatives of lasting importance, one in the face of active faculty opposition and the other after decades-long neglect. The first was his role in the decision in late 1993 to lay the groundwork for the creation of a department of biomedical engineering under the leadership of Van C. Mow, which will be discussed in the next chapter. The second was addressing the undergraduate recruitment and admission problem and establishing mechanisms that continue to pay dividends.[56]

ADMISSIONS TURNAROUND

The problem confronting deans since 1959, when SEAS reacquired the authority to recruit students directly from high school, was to do so in numbers large enough to help with the school's finances, while of a quality sufficient to satisfy the skeptical faculty in Hamilton Hall (Columbia College), who provided most of the instruction engineering undergraduates received in their first two years. In those three decades engineering managed its own recruitment and admissions. An unintended but long acknowledged consequence of this arrangement was that the two schools appeared to would-be applicants and college counselors to be competing with each other. Another was that the recruitment efforts of the better known College were many times greater than the engineering school's and its admissions numbers were consistently higher. In the late 1980s Columbia College averaged seven thousand applications a year, from which it produced its class of eight hundred; SEAS made its class of two hundred fifty from eleven hundred applications.[57]

From the early 1960s into the mid-1980s SEAS accepted between 50 and 60 percent of all applicants, while often having less than 40 percent of those accepted applicants later enroll. In the mid-1980s, a decade-long decline in the national college-age population, sharpest in the northeast where the SEAS applicant pool was concentrated, made a bad situation worse.[58]

In early 1991, Dean Auston convened his three-person admissions team to seek advice on how the school might recruit more and better undergraduates. One suggestion, offered by Joseph Ienuso, associate director of admissions, was to have SEAS and the College adopt a common application form and then move to merge their recruitment efforts. The benefits to SEAS in doing both were clear: take advantage of the College's greater name recognition; eliminate the appearance of competition; cut recruitment costs; increase applications. Those for the College were less apparent. When presented with this idea, admissions staffers in Hamilton Hall at first saw little in it for them.[59]

No question that the proposal for a common application came from SEAS and that the expected benefits all accrued to SEAS. The challenge for Dean Auston and his admissions staffers, Ienuso and later Eric Furda, was to convince Columbia College Dean Jack Greenberg and his admissions team that it would help the College increase its applications—what later came to be called the "Enhancement and Enlargement" initiative—to produce a larger entering class without an accompanying drop in quality.[60]

University officials had already been looking into the possibility of increasing undergraduate enrollments and tuition income to offset chronic budgetary shortfalls in the arts and sciences graduate programs. In the spring of 1990, University Provost Jonathan Cole contracted with Jan Krukowski Associates to produce a "Student Recruitment Marketing Plan" for both Columbia College and SEAS. The first installment would be an attitudinal survey of a sampling of high schoolers who had inquired about the applications process of either the College or SEAS but who subsequently did not apply.[61]

The survey revealed significant differences between the two schools. Location in New York City, for example, while a positive consideration, was more so in the thinking of possible College applicants than in those who considered applying to SEAS. Another finding was that Columbia's being an "Ivy" institution mattered more to those interested in the humanities and social sciences than to those planning to study science or engineering.[62]

The most revealing finding of the survey, though it should be no surprise here, was that Columbia was especially weak in attracting prospective science students, particularly those interested in the physical sciences and that strengthening the quality of the student body in both science and engineering was likely to require initiatives to improve the quality of Columbia's reputation in the sciences.[63]

Several marketing strategies flowed from these findings. One was that Columbia should emphasize New York City in its recruitment materials, not only its cosmopolitan and intellectual character (important to its traditional applicant pool) but opportunities it afforded for acquiring "hands-on" scientific and engineering learning experiences. Another was to emphasize features that distinguished Columbia from its Ivy peers' negative stereotypes (e.g., conservative, homogeneous), especially among those interested in science or engineering.[64]

But the most consequential recommendation was organizational:

Merge the admissions and financial aid offices, staffs, and operations of the College and the recruitment functions of SEAS. . . . From virtually every perspective—marketing strategy, cost-effectiveness of recruitment efforts, and actual recruitment outcomes—a merger of efforts makes sense. [65]

And so in 1992, at the direction of SEAS Dean Auston and Columbia College Dean Jack Greenberg, the admissions staff of the engineering school moved from Mudd to Hamilton Hall, longtime home of Columbia College. Lawrence Momo, a College staffer, became director of undergraduate admissions, SEAS's Joe Ienuso, director of technical services. When Momo left in 1996 to become director of admissions at the Trinity School, Eric Furda, who had come to Columbia in 1991 as an SEAS undergraduate recruiter and experienced firsthand the confusions of the bad old days, became acting director. The following year, Furda's first, was also the first SEAS class to top two thousand applications, more than double the number applying seven years before.[66]

Some of this increase can be attributed to improving demographic trends, some to an improving perception of New York City, and some should be credited to the unified approach to the marketing of Columbia's undergraduate schools. But credit is also due to Auston's persistence in getting the support of College and university officials for an arrangement that disproportionately benefited SEAS. The fact that SEAS admissions statistics from the mid-1990s have moved upwards hand-in-hand with those of Columbia College, and has been so noted by four successive deans of Columbia College, provides testimony to the central role engineers play in a resurgent Columbia.[67]

Auston's had been Columbia's shortest engineering deanship in the twentieth century and one that two decades later can claim few admirers among engineers who lived through it. Yet even among faculty in the department put most at risk by the dean's ambitious plans, it has been said of SEAS in the early 1990s, "even as [the] Auston airplane approached the cliff, the seeds of renewal were already discernible." A school on the long-term mend should be able, every now and again, to weather a dean who does not work out. Both school and dean went on to better times.[68]

9

A SCHOOL IN FULL 1995–2007

It's a good story.

—PROFESSOR NICHOLAS TURRO ON THE 1998–2000 RAPPROCHEMENT OF
THE DEPARTMENTS OF CHEMISTRY AND CHEMICAL ENGINEERING,
2012 INTERVIEW[1]

Zvi, Zvi, don't leave Columbia!

—SEAS SENIORS CHANT AT COLUMBIA COMMENCEMENT, MAY 2007[2]

DURING THE HALF-DOZEN years on either side of the millennial turn Columbia's school of engineering moved to consolidate and build upon the achievements of the previous two decades. Inspired leadership in the dean's office, an upturn in the demographics for college-agers, a supportive university administration, a flourishing national economy, a New York City on the upswing, and, perhaps most of all, enterprising faculty and ambitious students all conspired to put the school in a condition of institutional well-being and collective self-confidence not seen since its opening years. "At last," Professor of Applied Physics Gerald Navratil, who joined the faculty in 1978, said of the late 1990s at SEAS, "we had the wind at our backs."[3]

"A WONDERFUL RIDE": THE GALIL DEANSHIP

Unlike his earlier dark-horse candidacy for the chairmanship of the computer science department, Zvi Galil in 1995 was one of the front-runners to succeed David Auston as dean. Another leading candidate was Donald

Goldfarb, who since coming to Columbia from CCNY in 1982 had been a productive scholar and successful chair of the smaller industrial engineering operations research department. Goldfarb had also served as acting dean in 1994–1995. In choosing Galil, the search committee, Provost Cole, and President Rupp opted for the younger and more dynamic prospect, setting aside lingering doubts among some faculty about making one of the school's cocky computer scientists dean.[4]

Galil quickly allayed faculty misgivings by asking a skeptical member of the search committee, and the current vice dean, Civil Engineering Professor Morton Friedman, to serve as his vice dean. Though Friedman saw himself as an applied mathematician and thus on the "applied science" side of the applied science/engineering divide, by generation and department he was one with the "engineers." His knowledge of the university-at-large, acquired through his years of service on the University Senate, as well as a deep commitment to SEAS, made him a good complement to the school's relatively unknown dean. His twelve years as vice dean, embarked upon at age sixty-eight, Friedman later called "a wonderful and exhilarating ride."[5]

The Galil deanship commenced on the eve of the university's emergence from the straitened financial circumstances that had beset the end of the otherwise buoyant Sovern presidency. During those years, which encompassed the last two years of Gross's tenure and the entire Auston deanship, SEAS was feeling especially put upon by the standing financial arrangements with Low Library. In addition to the complaint common among Columbia's income-generating professional schools that they were subsidizing the arts and sciences, SEAS labored under the peculiar burden of an annual charge of $2 million to help pay down the debt incurred in the construction of Schapiro Hall, now more often referred to as CEPSR, for Center of Engineering and Physical Science Research. As Galil later described the feeling: "It was one thing to subsidize arts and sciences if you were rich like the medical and business school, but another for the engineering school which was itself broke."[6]

Help was on the way. The new university-wide financial arrangement that went into effect in 1994 allowed the professional schools to keep their income, whether from tuition, research overhead, or the new revenue stream of patent income. The schools would be charged an annual tax, in the case of SEAS during Galil's first year as dean, 57 percent of the school's current revenues. That meant that SEAS kept only 43 percent of the money it took in. But it also meant that if SEAS could increase revenues, it kept the whole increase. "I'm very big on providing incentives," Galil remarked. During his twelve years as dean, income from tuition, endowment, annual giving, distance-learning programs, and patents all increased markedly. The university's effective tax rate on SEAS dropped from 57 percent to 30 percent.[7]

Most of this additional money went to hiring new faculty. In 1996 the SEAS faculty had 92 members, about the number twenty-five years earlier; by 2007 it had 155 members, an increase of 75 percent. Each year of Galil's deanship the faculty grew by 5 or 6 members. The number of women on the faculty grew from 4 to 20. Unlike junior faculty hires in the hard-pressed 1970s and 1980s, when start-up funds were unavailable to tide them over until they could generate their own grants, by the late 1990s assistant professors came with funding commitments as high as $100,000. Their tenure prospects brightened accordingly.[8]

COMING OF AGE AT THE MILLENNIAL TURN: COMPUTER SCIENCE

We created a culture of internal competition for others to emulate.

—PROFESSOR OF COMPUTER SCIENCE, SALVATORE STOLFO[9]

The growing pains of the 1980s and the persistence of polarized conflict beyond Joseph Traub's founding years as chairman (1979–1989) earned computer science a schoolwide reputation for infighting. The tensions were still very much in evidence upon the arrival in the early 1990s of a second generation of junior faculty in the persons of Shree Nayar, Kenneth Ross, and Steven Nowick, all of whom eschewed the fray. Despite being a party to the earlier squabbling, Zvi Galil proved a mender and conciliator during his chairmanship (1989–1994). He was followed by Alfred Aho (1994–1997), who came to the department from Bell Labs in 1993, Kathleen McKeown (1997–2002), the first woman to chair an SEAS department, and Henning Schulzrinne (2002–2007), hired in 1996 as a joint appointment with electrical engineering. All these chairs, if not presiding over a veritable peaceable kingdom, kept the department focused on research and teaching.[10]

When Bell/AT&T Labs in New Jersey cut back on basic research in the early 1990s, SEAS was a principal beneficiary of the ensuing diaspora of its top scientists. No department more successfully "feasted on Bell Labs" than computer science, which, in addition to Aho (1997), secured the services of Julia Hirschberg (2002); Vladimir Vapnik (2003); Mihalis Yannakakis (2004), after an interim year at Stanford; and Steven Bellovin (2005). Some came because a move to Columbia did not require a change in residence, but all sensed they were joining a department—and a school—whose star was on the rise.[11]

Meanwhile, the department made a series of junior appointments, most of whom have since been tenured and now constitute the department's third generation. These included Luis Gravano (1997), Jason Nieh (2000), Stephen A. Edwards (2001), Vishal Misra (2001), Dan Rubinstein (2001), Peter Belhumeur (2002), Rocco Servedio (2002), Tony Jebara (2002), Tal Malkin (2003), Luca Carloni (2005), Eitan Grinspun (2005), and Angelos Keromytis (2006). "I think we've done remarkably well finding good really good people," a senior member said of the department's hiring policies at both the senior and junior levels. "We hire to raise the average."[12]

The intense student interest in computer science in the first decade of the department's existence almost instantly made it SEAS's largest instructional unit. It retained that distinction in the 1990s, drawing students from all four undergraduate schools, and from other professional schools. At home, however, the department encountered growing competition from biomedical engineering, which had particular appeal for women undergraduates, and industrial engineering/operations research and its financing engineering major. By the close of Galil's deanship, computer science had ceded its earlier status as the school's new kid on the block to still younger claimants, even as it acquired the respectability that sometimes comes with middle age.[13]

"A GOOD STORY": CHEMICAL ENGINEERING RESTORED

Several of the programmatic goals of the Galil years had been envisioned earlier in Dean Auston's ambitious reimagining of the school. It took a combination of an improving financial climate and Galil's less confrontational administrative approach to achieve them. Yet before any of these initiatives could be pushed, one from Auston's deanship had to be dismantled. The forced marriage between the Krumb School of Mines (aka department of mineral, metallurgical and mining engineering) and chemical engineering, which resulted in "The Henry Krumb School of Mines, Mineral Engineering and Materials Science," was undertaken to effect departmental consolidation and make broader use of the Krumb endowment. It produced neither of these organizational or financial outcomes. The six chemical engineers resented their minority status and bridled under the chairmanship of Ponisseril Somasundaran, who remained a strict constructionist in the use of the Krumb millions.[14]

Immediately upon becoming dean, Galil effected a separation, restoring the Krumb School faculty to their earlier departmental autonomy under the chairmanship of Nickolas Themelis, while placing the chemical engineers under the receivership of two distinguished members of the university's chemistry department, Professors George Flynn and Nicholas Turro. Two years later the chemical engineering department was back on its own, with renewed interest in extending its reach into new, well-funded areas of research. One such area involved human genomes and DNA sequencing, which with the appointment of Jingyue Ju in 2001, became an important part of the department's research agenda and a source of departmental solvency. The Cambridge-trained Ben O'Shaughnessy, whose research involves chemical engineering at the molecular level and the study of viruses that attack the human body, was hired in 1992 and secured tenure in 1998, the first in his department in eleven years. O'Shaugnessy's promotion was followed by that of Alan West, an electrochemical engineer hired in 1994 and tenured in 2000.[15]

The leadership issues that had earlier bedeviled the department were resolved with the recruitment of Jeffrey T. Koberstein, a specialist in polymer science, from the University of Connecticut and his assumption of the chairmanship. With his appointment, along with those of O'Shaughnessy and West, the department became significant participants in the increasing interchange between engineering and the health sciences, where for the previous twenty years Edward Leonard had been its sole point of contact. In 2005 Alan West became chair. He then proceeded to hire as a senior colleague Sanat Kumar, who had been at RPI, and whose research focuses on energy storage devices. He would later succeed West as chair. Turro and Flynn stayed on as active members of the department. In an interview shortly before his death in December 2012, Turro called the rapprochement between chemistry and chemical engineering, ninety years after their falling out, "a good story."[16]

THE GREENING OF MINES

Into the mid-1990s, the department of mining, minerals and metallurgy (aka the Krumb School of Mines) remained the oldest component of SEAS and arguably its most hidebound. The Krumb benefactions, dating back to 1958 and 1962, represented an ongoing endowment of $36 million committed to this single department and "to no other purpose." If Columbia violated these terms it would clear the way for the default beneficiaries, Sloan-Kettering and Lenox Hill Hospitals, to lay claim to half of what remained of the endowment. Thus, for three decades members of the Krumb School, who numbered twenty-four at one point in the early 1970s and occupied three entire floors of Mudd, enjoyed a budgetary independence and financial insulation unknown elsewhere at SEAS and likely the university. The department had had its share of distinguished faculty, including five NAE members, but few undergraduate majors and only a small contingent of graduate students. Faculty in other departments and a succession of deans concurred that the Krumb millions could be better spent elsewhere.[17]

Galil's undoing of Auston's merger left the six remaining School of Mines faculty for the moment a free-standing department, under the chairmanship of the reform-minded Professor Nicholas Themelis, to await further developments. It was by then clear to most department members, and certainly to the chairman, that at the end of the twentieth century a set of scholarly activities defined in the 1860s and organized around the needs of the mining and metallurgical industry no longer made sense. A new configuration was required if its engineers were to have a place in the school's future.[18]

An early sign of what that future might look like appeared in 1995 with the founding of the Earth Engineering Center, with NSF funding and under the direction of Nicholas Themelis. A clearer picture emerged from discussions the following year among Themelis, Dean Galil, Jeffrey Sachs, the second director of the newly established Columbia Earth Institute, the ubiquitous vice provost Michael Crow, and representatives of the Lamont-Doherty Earth Observatory. The upshot of these talks was that if the department were to have a future, it meant going green, that is, aligning its instruction and research with the sustainability concerns identified with the environmental movement. Even as this turnabout was being considered, in 1997 four department members of the School of Mines (Siu-Wai Chan, James Im, Gertrude Neumark, and Daniel Beshers), all working in the subfield of materials science, moved over to the newly formed department of applied physics/applied math. As Paul Duby, a thirty-year veteran said of his old department: "We were being hollowed out."[19]

The required metamorphosis proved less of a reach that one might think. Several past members of the department, among them the doughty Herbert H. Kellogg, while working closely with the mining industry, began in the 1970s to promote processes that reduced the toll mining and the metallurgical industry exacted on the environment. Research interests of other members, including Duby and Themelis, were of a similar environmentally cognizant character. In a sense, they had been doing environmental engineering without knowing it.[20]

In 2003 the Columbia trustees formally approved this reorientation by changing the department's name to the department of earth and environmental engineering. The scholarly conversion was further affirmed by the installation of its first chairman, Peter Schlosser, a specialist in the earth's hydrosphere, and since 1989 a professor at Lamont-Doherty. Schlosser was soon joined by two senior appointees, Upmanu Lall (2005), a specialist in water conservation, and Klaus Lackner, a specialist in efforts to reduced greenhouse emissions. Two junior appointments were also made: Marco Castaldi (2004), a specialist in the recycling of carbon dioxide; and Kartik Chandran (2005), who studies the influence of nitrogen on global climate. None had ties to the mining industry and none, unlike retired professor of mining engineering Colin C. Harris (1952–1990), risked disparagement by other Columbia scientists as SEAS's "professor of steel." With three hold-overs from the Krumb School days, Paul Duby, Ponisseril Somasundaran, and Tuncel Yegulalp, and only Yegulalp primarily identified with mining, the department had made a likely irreversible redirection of the department's teaching program and research agenda.[21]

Nor should the magnitude of the shift of the technical skills required be exaggerated. As Duby has suggested, the same scientific knowledge and engineering know-how once put in the service of the mining industry to maximize profit is equally applicable to limiting the adverse impact of human activity on the environment. Earlier research funding coming from International Nickel or US Steel or the Bureau of Mines might now be provided by the Environmental Protection Agency or NIH or the Ford Foundation. Jobs for SEAS graduates in mining or metals processing are now to be found at federal agencies, nonprofits, and private companies taking on the challenges of sustainability, of ensuring clean water and cleaner energy, of coping with climate change. Again, Professor Duby: "The underlying science is the same."[22]

A ROOM OF THEIR OWN: APPLIED PHYSICS/ APPLIED MATH

Efforts to gather the school's departmentally dispersed applied physicists and applied mathematicians, most of the former initially in mechanical engineering, the latter in civil engineering, date back to the early 1960s and the collaborative undertakings of the applied physicist Robert A. Gross and the applied mathematician C. K [John] Chu. Their training—both were PhDs in departments where many faculty were not—and research interests aligned them less with traditional engineers of their departments than with the physicists and mathematicians in what was still the faculty of pure science. Something of a halfway house was formed in the early 1970s when Gross and Chu had themselves designated "professor of engineering science" (there was no such department). This was followed in 1978 by the creation of a department of applied physics and nuclear engineering, with Gross as its founding chairman. When the last of the faculty identified with nuclear engineering retired, the department in 1989 quietly dropped the second half of its name. By then the department had added to its faculty ranks the applied physicists Irving P. Herman (1984), Michael Maul (1985), and Wen Wang (1987). But not until 1998 was a department of applied physics and applied mathematics (APAM) authorized by the university trustees. This followed directly on a financial windfall to SEAS of epic proportions.[23]

In 1996, the Chinese businessman Z. Y. Fu, founder of the Tokyo-based Sansiao Trading Company and brother-in-law of applied mathematician C. K. "John" Chu, gave Columbia's School of Engineering and Applied Science a gift of $26 million. This was the largest single gift ever given to Columbia by a single benefactor. Dean Galil, who had a minor role in securing the gift ("this was John and his sister's doing"), announced that the school would henceforth bear the name (its fifth) "The Fu Foundation School of Engineering and Applied Science of Columbia University." He further announced, in keeping with the wishes of the donor and his brother-in-law, that the money would go primarily to support the departments of electrical engineering and computer science, the department-to-be of biomedical engineering, and the newly constituted department of applied physics and applied mathematics.[24]

Money alone likely would not have brought the APAM department into being. Endorsement, or at least acquiescence, by the cognate arts and sciences departments of mathematics and physics was essential. That the physicists readily agreed speaks to the improving relations SEAS faculty had developed with them by the 1980s, thanks in considerable measure to

FIGURE 9.1 Z. Y. Fu, George Rupp, and Zvi Galil (l to r). Columbia's eighteenth president, George Rupp (1994–2003), and SEAS's eleventh dean, Zvi Galil (1996–2007), honoring Chinese businessman Z. Y. Fu (1919–2011), whose 1997 gift of $26 million led to the renaming of the school as the Fu Foundation School of Engineering and Applied Science.

Source: SEAS Communications.

the ambassadorial efforts of Robert Gross and Richard Osgood. That the mathematicians went along with a department of applied mathematics in the engineering school was a near thing, but they did. Current members of the mathematics department remember their senior colleagues once regularly berating their engineering contemporaries with such value-laden distinctions as "there is good mathematics and there is applied mathematics." But times were changing.[25]

The subsequent growth of the applied physics and applied math department during the Galil deanship took several forms. The first occurred in 1998 when Gertrude Neumark, Siu-Wai Chan, James Im, and Dan Beshers left the Krumb School of Mines with Dean Galil's blessing to become the materials science component of APAM. Neumark retired in 2000 and

Beshers in 2005, although he continued teaching, while Chan, the department's first woman faculty member, and Im, who studies silicon at the nanoscale level and works with polycrystalline materials, stayed on. The second was through a series of junior appointments, among them Chris Wiggins (2000) in applied math and William Bailey (2001) in materials science. Faculty growth also occurred through agreements with the newly organized department of earth and environmental science (EES) in the faculty of arts and sciences (again brokered by Michael Crow) and with the SEAS department of earth and environmental engineering (EEE) that made possible the joint appointments of Lorenzo Polvani (1997), Adam Sobel (2000), and Marc Spiegelman (2003). Another deal with the Columbia math department produced the joint appointment of the applied mathematician Michael Weinstein (2002).[26]

But surely the most spectacular form APAM departmental growth took involved the joint appointments with the Columbia physics department in 1998 of the Bell Lab applied physicists Horst Stormer and Aron Pinczuk, as worked out over two years by physics chair Norman Christ and APAM chair Gerald Navratil, with funding from Columbia brokered by Michael Crow and the Fu Foundation–enriched SEAS. Pinczuk's research is in condensed matter physics and focuses on understanding the properties of novel materials on the nanoscale. Nine months after Stormer's appointment he won the Nobel Prize, which he shared with two collaborators, for the discovery of a new form of quantum fluid with fractionally charged excitations. On the day of the announcement the department secretary posted a note on his classroom door: "Class cancelled, instructor won Nobel Prize."[27]

"The Stormer appointment," Michael Maul recalled fifteen years later, "changed everything," even what SEAS professors wore on campus and to class, their Nobel laureate colleague's chinos-or-dungarees-and-open-collar-or-tee-shirt style suddenly the norm. More substantively, his "disdain of departmental structures, seeing them as inhibiting faculty from getting work done," and of the endless administrative deliberations, which likely contributed to his later decision to take early retirement, Maul remembers, "was contagious." During his dozen years at Columbia, years of enthusiastic identification with the APAM department and SEAS, the once almost automatic disdain evidenced among other parts of the university when contemplating collaborative arrangements with the university's engineers, and the engineers' earlier diffidence in entering into these arrangements, were finally put to rest.[28]

A SECOND CHANCE FOR BIOMEDICAL ENGINEERING

Central to effecting interschool collaborations in the mid-1990s was the new emphasis placed on them by incoming President George Rupp. A specialist in Eastern religions, Rupp had experience working with engineers as president of the technology-centered Rice University. He brought to Columbia a focus on undergraduates and a determination to bring about interdepartmental and cross-schools collaboration in teaching and research. It was likely with an eye to the latter that he kept on as university provost the Sovern-appointed Jonathan Cole and endorsed the ongoing matchmaking work of Vice Provost Michael Crow. Both continued putting revenues derived from patents into new areas of research that had their own income-producing potential.[29]

A prime beneficiary of this entrepreneurial surge was biomedical engineering. When Van C. Mow and W. Michael Lai accepted joint appointments as professors of orthopedic surgery and mechanical engineering back in 1986, Mow had hung his hat at his laboratory in Harkness Pavilion up at the medical school while Lai made a place for himself in the mechanical engineering department in Mudd. There he was joined in 1988 by Gerard Ateshian, a Columbia PhD who had been a student of Mow's. In 1991 Mow was elected to the National Academy of Engineering.[30]

On December 30, 1994, Mike Crow and Dean Auston came calling on Mow. They were ready to move on the creation of a joint biomedical engineering program as a first step toward establishing a department, to be based at SEAS but to include faculty at the medical school. They wanted Mow to head it. This was to be one of the first major undertakings of President's Rupp's interschool collaborative initiative as detailed in his administration's

"Strategic Planning Document." Start-up funding would be provided by the university, which would allow Mow and his SEAS colleagues, among them Edward Leonard, to put together a grant proposal to the Whitaker Foundation, then a major philanthropic force in bioengineering. Mow, faced with a looming shrinkage in funding to pay for more than forty staff working in his uptown lab, expressed interest.[31]

The stiff terms Mow set for a Mechanical Engineering and Orthopedic Surgery Program:

—the mechanical engineering department endorse the program (made easier by Deresiewicz and Freudenstein having retired and Chevray gone);

—the university match with three dollars in support every one dollar Mow raised;

—SEAS provide 50,000 square feet of laboratory space on the Morningside campus;

—the new Mechanical Engineering and Orthopedic Surgery Program move as quickly as funding and staffing allowed to become a free-standing department of biomedical engineering, with Mow as chairman;

—the target size for the department be twenty members.

When apprized of these terms, Provost Cole called and asked Mow, "Are you serious?" When told he was, Cole responded, "OK, you're on."[32]

The first step to department status was taken later in 1995 with the creation of a joint SEAS/P & S Center for Biomedical Engineering. The second was authorizing, in anticipation of the department's creation, appointments in three specialties agreed upon with the Whitaker Foundation, which provided $1.5 million in planning funds: X. Edward Guo (1996), a specialist in biomechanics, as assistant professor of mechanical engineering; Andrew Laine (1997), a specialist in medical imaging, as associate professor of mechanical engineering; Kevin D. Costa (1997) and Clark Hung (1997), both specialists in cellular and tissue engineering, as assistant professors of mechanical engineering. Similar temporary arrangements were made with

FIGURE 9.2 Van C. Mow (1938–), born in China, came to Columbia in 1986 from RPI, where he had been a student and professor. Founding chairman of SEAS Department of biomedical engineering (2000–2012).

Source: SEAS Communications.

electrical engineering to secure the appointment of Paul Sajda (1999), a specialist in medical imaging, with the expectation that he would soon be joining Mow, Guo, Laine, Hung, and Costa in a department of their own. [33]

With Galil and Mow now working in tandem after earlier scuffles, a $7.5 million Whitaker Foundation start-up grant in hand, three-to-one matching university funding as arranged by Mike Crow, office space carved out of the engineering library on the third floor of Mudd, a core faculty in place and additional lines authorized, preparations were complete. On January 1, 2000, the university formally announced the creation of an eight-person department of biomedical engineering based in SEAS and with Van C. Mow its chairman. Two years earlier he had been elected to the Institute of Medicine, the only member of the university to hold memberships in two National Academies.[34]

Between 2000 and 2006 the biomedical engineering department made eleven more appointments at the assistant professorship level, five to women, distributed among its three areas of departmental specialization. All but two have since received tenure. By 2006 biomedical engineering had established itself not only within SEAS but through its many links with departments in the Arts and Sciences and in health sciences as a university-wide presence.[35]

THE QUANTS ARE COMING

Back in 1992 when Dean David Auston cast about SEAS's departments looking for those to consolidate or eliminate, the eight-person industrial engineering and operations research department clearly qualified. Although the operations research side continued to attract students, often some of the school's best undergraduates, including earlier SEAS's two Nobel laureates, Robert C. Merton (OR 1966) and Alvin Roth (OR 1971), industrial engineering held little appeal. Undergraduates had no interest in manufacturing, while the earlier flow of sponsored graduate students supplied by IBM and AT&T had dried up. Moreover, the kinds of research industrial engineers and those in operations research traditionally did, usually in the profit sector, federal agencies were disinclined to fund.[36]

The senior appointment in 1982 of Donald Goldfarb, a 1966 Princeton PhD in chemical engineering who founded the computer science department at City College, followed on the death of Edward Ignall. Shortly after arriving, Goldfarb became chairman, but the department remained home to several other senior faculty, among them Morton Klein and Cyrus Derman, who had led the department in the postwar era but had since ceased doing research. Seymour Melman remained the university's most vocal critic of the defense establishment but contributed little to the department's core offerings. None of the senior faculty had sponsored funding. Goldfarb's subsequent characterization of the department he inherited: "moribund."[37]

David D. Yao, born in China and a recent PhD in operations research from the University of Toronto, was appointed assistant professor in 1983. His prospects and those of contemporaries Michael Pinedo and Zvi Schechner were not promising. As with several other engineering departments in the 1970s and early 1980s, IE/OR had been unable to secure tenure for several promising assistant professors. In some cases, aware of the situation, junior faculty simply left. Three years after his arrival, Yao accepted an offer of an associate professorship from Harvard's Division of Applied Science and left. Two years later, he returned to Columbia as a professor, but to a department still mired in "a no-growth situation" and still with no real plan for breaking out of it. He then joined forces with two recent junior appointees, Karl Sigman and Guillermo Gallego, to confront the question, "What should we be doing now?" The answer, Yao recalled in 2011, "became pretty obvious. We connect with New York City's financial services community."[38]

Some of what made the answer obvious was that several recent IE/OR graduates had already found their way to Wall Street, where their analytical and program-writing skills put them in demand. The views of these alums and current students interning on Wall Street were actively solicited by Karl Sigman. The feedback: If the department could supplement traditional OR courses with others focused on the financial services industry, it would be in business.[39]

This did not happen overnight. None of the three faculty centrally involved—Yao, Sigman, or Gallego—had previously conducted research on financial markets. Each had to be retrained in finance before he could hope to teach it. In Gallego's case, retraining occurred during his 1996–1997 sabbatical at the Stanford Business School; in Yao's, a year of teaching in Hong Kong. Still, the effort to get up to speed on financial markets prompted Sigman to wonder, "How long, if ever, before I can write a paper on this?" Yao later minimized the problem: "As engineering faculty we always have to reinvent ourselves."[40]

In 1997–1998, two new graduate courses—"OR—Methods in Finance, I," taught by Gallego, and "OR—Methods in Finance, II," by Yao—were added to the department's offerings. The following year the courses were repeated, with "Methods in Finance" replaced by the new term of choice,

"Financial Engineering." In 1998 Gallego and Sigman became codirectors of the Financial Engineering Program, with Sigman becoming director in 2001. The following year, the department hired Emanuel Derman, a Columbia PhD in applied physics who had spent nearly two decades at Solomon Brothers and Goldman Sachs, where he headed their quantitative strategies group. By then financial engineering had become the school's most popular major and the school became a major supplier of quantitatively sophisticated personnel to Wall Street. That so many of SEAS's best students in the years around 2000 were going immediately into the financial services sector upon graduation caused the IE/OR department to see the need for a broader major, involving engineering management systems, that would draw on faculty expertise in pricing, queuing, inventory, and the provision of health services.[41]

In the course of "reinventing itself" in the early 2000s, the IE/OR department came up with a novel business plan for expansion. It did so by developing professional MS programs that focused on training students who had an interest in directly joining the technical workforce rather than go on for a PhD. This required the department to focus on students. New York City was ideal to start such MS programs given the demand for such students in the local economy. During the chairmanship of Guillermo Gallego (2002–2008) an overhaul of the department's staffing structure resulted in greater attention paid to students, both undergraduate and graduate, from orientation to graduation. Also included was a careful monitoring of classroom teaching by new assistant professors and the department's many adjuncts, as well as properly acknowledging senior faculty for their commitment to teaching. The department began providing direct assistance with student job hunting, encouraging human resource departments at banks and brokerages previously focused on MBAs to consider hiring engineers. As Gallego put it, once the department "got the first batch out there," those graduates made the way easier for batches that have followed.[42]

Also required was administrative agreement that IE/OR's strategy of funding faculty hiring through expansion of the MS programs served both the department and SEAS. This Dean Galil provided in 2001 by allowing the department to proceed with two senior hires, those of Clifford S. Stein from Dartmouth (jointly with computer science), and a year later Ward Whitt from Bell Labs. Stein is the coauthor of *Introduction to Algorithms*, a best-selling textbook in algorithms that has been translated into fifteen languages. He has also served as department chair. Whitt was elected to the National Academy of Engineering in 1995 for "advances in understanding and analyzing complex queues and queuing networks, leading to improved telecommunications systems." The department's turn-of-the-millennium revival occurred in time to allow faculty to provide the services to industry broadly with mathematically based marketing and delivery strategies, even as its graduates took full advantage of local employment opportunities provided by the booming New York City–centered financial services sector.[43]

AT THE MILLENNIAL TURN: ELECTRICAL ENGINEERING

Most of the growth early in the Galil deanship occurred in the school's newest departments. But the school's increasingly prosperous circumstances made it possible to strengthen all nine departments, including what outsiders at their risk call the "mature" disciplines of electrical, civil, and mechanical engineering.

Of all the engineering school's departments, electrical engineering has been least subject to the kinds of temporal swings that characterized the history of other departments. The third to be formally organized, in 1889, it has consistently been viewed as one of the school's strongest departments and the one most closely aligned with its university counterpart, the physics department. Early faculty included such internationally known scientists as Michael Pupin, John Harold Morecroft, and Edwin Armstrong. Faculty of the immediate postwar period such as John Ralph Ragazzini, Lotfi A. Zadeh, and Cyril M. Harris sustained its reputation as a locus of engineering research. The department has provided the school with two deans, Joseph W. Barker (1930–1946) and David H. Auston (1991–1994), as well as Ralph Schwarz who twice (1975–1976, 1990–1991) served as interim dean. Electrical engineering also enjoyed a well-earned reputation as being the most intellectually demanding of the school's undergraduate programs, as well as one of the more heavily enrolled. Its graduates include the aeronautical pioneer Grover Loening (1908), Armstrong (1913), Eliahu Jury (DES 1953), who developed the advanced Z-transform used in digital control systems, and Rudolf Kalman (PhD 1957), the coinventor and developer of the Kalman filter, a mathematical algorithm used in digital signal processing.[44]

For all that, electrical engineering has had its setbacks. It unsuccessfully opposed the creation of a separate computer science department in the 1970s and a separate biomedical engineering department in the 1990s, arguing in the first case that computing properly fell within its purview, and in the second, that faculty resources were being spread too thinly. Still, on balance, it survived the rough 1970s and 1980s in relatively good order, thanks in large part to two strategic senior appointments, Mischa Schwartz (1973) and Richard M. Osgood, Jr. (1982), and several junior appointments, among them Yannis Tsividis (1978), Dimitris Anastassiou (1983), Aurel A. Lazar (1983), and Charles Zukowski (1985), all of whom have made their careers at Columbia. Indeed the department's success in appointing junior faculty who went on to secure tenure likely exceeded that of any other department.[45]

Securing $20 million in NSF funding for the Telecommunications Research Center, the 1983 brainchild of Schwartz and then department chair Thomas Stern, which became SEAS's biggest underwriter of faculty and PhDs in the late 1980s and early 1990s, assured the department continued administrative favor, as did Anastassiou's lucrative (to SEAS as well as him) licensing of his invention of an algorithm for compressing video signals. EE was the only traditional department included by Dean Galil to be among the direct recipients of the Fu Foundation largesse.[46]

Electrical engineering also led the way in the mid-1990s with a series of what were then for the school still novel joint appointments: in 1995, Tony Heinz, a senior joint appointment with the Columbia physics department; in 1996, Henning G. Schulzrinne, a joint appointment with computer science. That same year, Kenneth L. Shepard joined the department and would later accept a joint appointment in biomedical engineering. In 1999 Charles Zukowski became chairman; two years later the department hired its first woman faculty member, Keren Bergman, whose research centers on photonics, the science and technology of generating and controlling photons, as applied to the Internet traffic management system. In 2012 she became the second woman to chair an SEAS department.[47]

TO THE BRINK AND BACK: CIVIL AND MECHANICAL ENGINEERING

Two other of the engineering school's "mature" departments, mechanical and civil engineering, each at various points in the twentieth century arguably its most distinguished, became near casualties of the 1990s. Like chemical engineering, mechanical engineering went into receivership in the mid-1990s after three successive deans despaired of what they saw as the department's recalcitrance in the face of needed change. In the late 1990s the department functioned primarily as a boarding house for soon-to-be-gone biomedical engineers. SEAS's civil engineers avoided the imposition of leadership from outside their ranks, but not a sharp drop in faculty size, brought on by retirements and failed tenure cases, to half what it had been at the apex of their "golden age" of precomputational engineering mechanics in the late 1950s and 1960s. It also attracted fewer undergraduates than it once had. Both departments appeared on Dean Auston's "consolidate-or-close" list in 1992; five years later, neither appeared on Dean Galil's list of beneficiaries of the $26 million Fu Foundation largesse.[48]

Nor did mechanical or civil engineering attract the interest of Mike Crow during his canvass of the university for would-be collaborationists in new research ventures. The feeling was mutual. As one of the civil engineering old guard put it, Crow was "taking Columbia from a community of scholars to a bogus corporation," with his departmental solicitations "the beginning of the baloney."[49]

Yet both departments survived. Mechanical engineering benefited from leadership imposed from outside in 1994–1995, in the persons of SEAS professors John Chu and Gerald Navratil, but also from the steadying influence of the department's own W. Michael Lai, who chaired the department from 1995 to 2001. Lai was followed by Vijay Modi, whose tenure promotion in 1994 had been the first in the department in eighteen years. Civil engineering's turnaround, like its downward slide, was less dramatic, but was similarly facilitated by the quietly effective leadership of Rene Testa as department chair in the late 1990s.[50]

One sign of improving times in both departments at the century's turn was their renewed capacity to see junior appointments through to tenure. In mechanical engineering, Gerard Ateshian, who works on the growth and lubrication of human cartilage, secured tenure in 1998, followed two years later by Y. Lawrence Yao, whose research is in nontraditional manufacturing. In civil engineering, Raimondo Betti, the department's only hire in a decade when he came in 1992, was tenured in 1998, the first in the department in nineteen years. He was followed by Hoe I. Ling, hired in 1998 and tenured in 2005. By then, both departments had their tenured ranks restaffed, their leadership in good hands, and a green light from Dean Galil to recruit new faculty, especially those whose coming to SEAS would move the school into newer, more fundable, and more student-attracting specialties within their traditional fields.[51]

In civil engineering, the dean's go-ahead allowed the hiring in 2000 of Andrew W. Smyth, whose research centered on the monitoring of urban infrastructures, and two years later, Patricia J. Culligan, a specialist in geoenvironmental engineering, as associate professor and the department's first woman faculty member. That same year, 2002, a Columbia-trained PhD, George Deodatis, doing research on infrastructure risk analysis, left Princeton to join the department, where he had been a graduate student fifteen years earlier, as a professor. All three appointments gave the department strength where it most needed it, in computational engineering mechanics.[52]

At the start of the new millennium, mechanical engineering made six junior appointments in rapid succession. The first occurred in 2000, of Luc Fréchette, the first department hire whose primary research interest was in microscale science and technology. He was followed by Jeffrey W. Kysar, a Harvard PhD then at Brown doing research on the mechanical behavior of materials. Next came James Hone, a Berkeley PhD working on nanoelectromechanical systems (NEMS); in January 2003, Chee Wei Wong, a nanostructural engineer with a PhD from MIT; in 2004, Qiao Lin, then conducting research in microelectromechanical systems (MEMS); in 2005, Nabil Simaan, whose research was in medical robotics. All these hires in two departments only recently thought to have outlived their usefulness, and the publicity attending them, led to upswings in enrollments and majors in both departments. All boats lifted.[53]

"ZVI, ZVI, DON'T LEAVE COLUMBIA!"

Galil's twelve-year deanship benefited from an extended era of rising fortunes of the university overall that encompassed the presidency of George Rupp (1994–2003) and the first three years of the Lee Bollinger presidency. One of the most marked university developments of the period, and a specific goal of the Rupp presidency, was increased attention to undergraduate education. Here, SEAS was positioned to be both an outsized beneficiary of this initiative and Galil a conspicuous player.[54]

The first year of Galil's deanship SEAS accepted almost half of all undergraduate applicants, with only 30 percent of those accepted deciding to come. SAT scores ran 100 points below those of applicants accepted to the College. But by then both the improving college-age demographics and the benefits of the merged admissions process had begun to take hold. Thereafter, SEAS applications rose every year and admissions became steadily more selective (SATs of entering classes rose from 1,310 in 1996 to 1,480 in 2007 [out of a possible 1,600], fifth highest among American universities). By 2007 the school accepted only one in five applicants and saw its yield (percentage of admits registering) rise to 45 percent. In the same twelve-year span Columbia College moved from twelfth place among the most competitively selective colleges to being tied for fourth.[55]

Increased selectivity contributed to improved morale among engineering undergraduates. But surely some of their sense of well-being can be attributed to attention paid them by their dean. Galil was the first engineering school dean to achieve positive visibility among the university's nonengineering undergraduates and parity with the heads of the three other undergraduate units of the university in terms of student press coverage. His seeming ubiquity was further enhanced by students using a life-size cut up of him at meetings when other obligations precluded his presence. It mattered to engineering undergraduates that they mattered to the dean and that, university-wide, their dean mattered.[56]

FIGURE 9.3 Zvi Galil (1947–), with poster; born in Israel, came to Columbia in 1982 as professor of computer science; chair of department (1989–1994) and SEAS's eleventh dean (1995–2007); later president of University of Tel Aviv (2007–2009) and dean of the Georgia Institute of Technology College of Computing (2010–)

Source: SEAS Communications.

An example of Galil taking undergraduates seriously related to teaching performance. Course evaluations at Columbia date back to the early 1970s with the annual publication of the student-run and once avidly read *Barnard-Columbia Course Guide*. By the mid-1990s individual schools had developed their own instruments for securing student assessments of their courses. The question remained, however, if anyone other than students choosing their next semester's program or the occasionally maligned instructor took these evaluations seriously. When Galil began having conversations with SEAS faculty with poor ratings, then setting up a workshop for improving classroom effectiveness, both students and faculty had the answer. "I didn't want faculty to be spending all their time on their teaching," Galil later recalled, "but students have a right to expect, and the dean to require, that ineffective teachers take the time to improve their teaching." Meanwhile, increased prominence was given to teaching by the Engineering Alumni Association, which made two awards at Class Day, supplementing "The Great Teacher Award" bestowed annually by the Society of Columbia Graduates, which also recognizes engineering faculty. Almost gone were the days when engineering students regularly complained that their instructors could not be understood in class or approached outside of it.[57]

Alumni giving provides another indicator of "customer satisfaction." Between 1996 and 2007 the SEAS annual fund quadrupled from $600,000 to $2.7 million. The school endowment during this same twelve-year period more than tripled, from $125 million to $400 million. Favorable market considerations contributed to this growth, but so did the school's success in raising endowment dollars. The Class of 2007 achieved 100 percent participation in the Senior Class Gift, a new record.[58]

It was these same SEAS seniors who learned in early 2007 of Galil's decision to accept the presidency of the University of Tel Aviv, which his family had helped found and where Galil had his first teaching job. At the university commencement in May they took up the chant, "Zvi, Zvi, don't leave Columbia." SEAS faculty, fellow administrators, and his bosses in Low Library joined in. But leave he did, for what turned out to be a short-lived and controversial presidency, and a subsequent return to America as dean of the College of Computing at Georgia Tech.[59]

The twelve-year Galil deanship ranks as the most successful in the school's then 143-year history. Of Galil's dozen predecessors, only the first, Charles Frederick Chandler (1864–1897), and the tenth, Peter Likins (1976–1980), bear comparison. For all the energy and distinction Chandler brought to the School of Mines, his long deanship was marred at its end by conflicts with both the trustees and President Low, by which time the school had already entered into an extended period of intrauniversity marginality and declining national visibility. Likins may have been a more consequential dean than Galil, insofar as his deanship helped bring about the positive turnaround in the school's trajectory that has since been sustained by his five successors. Yet Likins was able to do only so much in four years. His successful efforts to return computer science to Columbia constitute an accomplishment equal to that of Galil's helping the biomedical engineering department come into being, but Galil was able to accomplish more in twelve of the university's "fat years" than Likins in his four "lean" ones.

THEMATIC CODA II: MARTHA OUT OF THE KITCHEN

I'm a simple man. I've never been able to understand the delineation between science and technology and engineering and innovation and all that stuff. To me it is all one thing.

—ALAN LESHNER, CEO OF AMERICAN ASSOCIATION FOR
THE ADVANCEMENT OF SCIENCE, 2002[60]

A development of great importance of the last forty years in the history of SEAS, Columbia, and American culture involves the waning of the earlier disposition to privilege the claims to institutional support and public approbation of "disinterested" science over those of problem-oriented engineering and applied science. Paul Forman, a historian of science and past curator of the Division of Medicine and Science at the National Museum of American History, has described the change as "the abrupt reversal of culturally ascribed primacy in the science–technology relationship—namely, from the primacy of science relative to technology prior to circa 1980, to the primacy of technology relative to science." Where technology was earlier considered subsidiary to and derivative of science, science is now viewed as just another form of technology.[61]

That Columbia fully subscribed to this earlier view of its engineers and applied scientists as intellectually beholden to and institutionally subordinate to its "pure" scientists has been a recurrent theme here. In Forman's accounting of this persuasion's ascendancy throughout the industrialized West, Columbians figure prominently among his articulators of this conceit. If Burgess goes unmentioned, the Columbia philosopher John Dewey, the Columbia engineering school alumnus Gano Dunn, the Columbia physicist I. I. Rabi, the Columbia literary critic Lionel Trilling, and the one-time Columbia-based sociologist Daniel Bell are all called upon to affirm the cultural primacy of science over technology in their day.[62]

So what happened to reverse this widely held cultural dictum? And what happened at Columbia, in the last four decades, to effect, if not the full reversal of places that Forman and others believe to have occurred, sufficient change to give the engineers and applied scientists equal standing with the scientists?

The departure of several of postwar Columbia's leading technology skeptics from the scene was one factor. Rabi was less influential after retiring in 1968, although remaining a presence on campus for another two decades (he died in 1988). Some who knew him in his retirement have suggested that he may also have softened his earlier views vis-à-vis applied science. Barzun left in 1969 for Texas; Bell decamped that same year to go to Harvard; Polykarp Kusch left Columbia in 1972, also for Texas; Trilling died in 1975. Within the space of a half-dozen years, the campus had lost its most articulate upholders of the primacy-of-science doctrine. No one of similar stature came forward to take their place.[63]

Changes in university leadership beginning in the early 1970s constitute another significant factor. Between 1889 and 1970, Columbia had been presided over by a civic leader (Low), a philosopher (Butler), a general (Ike), and a political scientist (Kirk). None was a scientist and none expressed anything but passing curiosity about the sciences, "pure" or "applied." Neither did any question the conventional wisdom favoring the former over the latter. The appointment of William J. McGill, a mathematical psychologist, who as a Columbia professor had been part of an interdisciplinary group of quantitative social scientists led by statistical sociologist Paul Lazarsfeld and members of the engineering faculty, broke the eight-decade-long pattern. Although preoccupied with putting Columbia's finances in order, McGill understood the university he presided over. "Harvard [and MIT] are about nature," he quickly concluded on arriving in 1956, "Columbia is about books." As president he invested both time and resources in upholding the place of science, and technology, on Morningside Heights.[64]

A similar shift occurred among the university's chief academic officers. The first in modern times to hold the position, though not the title of provost, was the redoubtable technology keptic John W. Burgess, a political scientist, who served as the senior dean of the three graduate faculties (1892–1904). The first provost so designated was William H. Carpenter (1911–1927), a specialist in German literature, followed by seven successive nonscientists.[65] Not until McGill made Polykarp Kusch his provost in early 1971 was the position filled by a scientist, albeit one known to be dismissive of Columbia's engineers. It was Sovern, himself a presidential departure, coming from a professional school, who broke the pattern by first appointing Dean of Engineering Pete Likins as co-provost in 1980, and then in 1982, Robert F. Goldberger, a medical research scientist and member of the Faculty of Health Sciences (1982–1989), as provost. Although often critical of the engineering faculty during his annual appearance at the school's spring faculty meetings, Goldberger is remembered by veteran faculty as taking a keen interest in the teaching and research activities of his engineering colleagues.[66]

By most reports Goldberger's successor as provost, the sociologist of science Jonathan R. Cole (1989–2003), was viewed by his engineering faculty in the "tough love" tradition of McGill/Sovern/Likins/Goldberger. His subsequent critique of Vannevar Bush's influential *Science—The Endless Frontier* (1945) in *The Great American University* (2009), attributing to Bush "the false dichotomy between basic and applied research that . . . burrowed its way into the psyche of the scientific community and created an imprecise model of the process of discovery and innovation," puts him firmly in the protechnology camp.[67]

Again, strong advocacy on behalf of Columbia's engineers and applied scientists by engineering deans Likins, Gross, and Galil contributed importantly to the enhanced standing of the school—in the university. It was duly noted by university officials during Likins's four years as dean that SEAS contributed significantly to the university's financial well-being by substantially increasing enrollments while restraining faculty growth, even as most arts and sciences departments experienced declining enrollments and ran deficits. Even while expanding his faculty, Galil earned SEAS similar recognition as more than paying its way.[68]

Negative developments affecting Columbia's scientists and the standing of their departments also altered SEAS's relationship with the university. Into the 1970s Columbia's greatest strength in the sciences was mostly in the physical sciences, specifically in physics and chemistry, along with mathematics. Despite its having a world-class medical school, and an earlier preeminence in genetics, postwar Columbia was not strong in the biological sciences. That gap in distinguished coverage mattered less during the Cold War when funding for physicists and chemists doing research even tangentially related to national defense was widely available; it came to matter far more when the Cold War wound down and defense-related funding dried up. Columbia found itself in the 1970s and 1980s scrambling to strengthen the biological sciences, not least because, especially in the post–Berlin Wall era, that's where the research money was.[69]

Columbia's engineers had also been recipients of Cold War–related funding and with their colleagues in physics and chemistry felt the pinch when it dried up. But as engineers and applied scientists, "mission-driven" as their critics once dismissively characterized them, they more handily than the university's supposedly "curiosity-driven" physicists and chemists adapted their research agendas to align with the new focus on biology and the health sciences. Whereas in the 1960s SEAS faculty pursuing research relating to health could be counted on one hand, by 2001 there were more than three dozen doing so.[70]

This "biological turn" has been good for Columbia's engineers in other ways. It has made them more visible on campus, more likely to enter into collaborations with welcoming colleagues in the biological sciences on the Morningside campus and at the medical school, and more fundable. Apropos this last point of increased access to research support, Thomas Cech, president of the Howard Hughes Medical Institute, a leading funder of medical research, had this to say in 2005 on the science–technology dichotomy: "We scientists like to think that our smart ideas are what are driving science forward, but we keep realizing that . . . the limiting factor really is . . . the technology." But for now, the penultimate word on the subject might be that offered in 2002 by the then CEO of the American Association for the Advancement of Science (the same organization to which Henry Rowland 130 years earlier delivered "A Plea for Pure Science"): "I'm a simple man. I've never been able to understand the delineation between science and technology and engineering and innovation and all that stuff. To me it is all one thing." Rosalind Williams noticed the same relative upgrading of technology vis-à-vis science among her MIT colleagues and Cambridge neighbors:

> A striking characteristic of the 1990s was the extent to which both personal and historical experience became dominated by technology. . . . Suddenly, it seemed, dinner conversations with friends began veering toward computers, what games your kids were playing, what software you were using, and soon what you were doing on the Internet.[71]

The last word, to bring the matter back to Columbia, must go to a Columbian. The earlier quote about "a false dichotomy" from ex-provost Jonathan Cole might serve, but remarks by Professor of Psychology Eugene Galanter, addressing a gathering of Columbians in 2003 on the subject "The Sciences at Columbia," makes the case for scientist–engineer parity more pointedly. C. P. Snow's famous *The Two Cultures* (1959), the idea for which Snow credited I. I. Rabi, serves as Galanter's point of departure:

> The two cultures are really at least three cultures. The redemption of science in the eyes of the humanist is made possible by scientists handing

over a sacrifice at the altar of intellectual superiority: the engineer. This fits comfortably with the general view that engineering is somehow a derivative of scientific knowledge. The conceit is that science formulates laws and theories (the deep stuff), and then engineers take these laws and use them to construct dongles and dohickeys.

Warming to his subject:

Scientists can in no way disregard the work of the engineer, and my reasons for this view are not that scientific advances depend upon engineering tricks, but rather that engineering is the contact point between science and humanity.[72]

Martha has left the kitchen.

10

A LEVER LONG ENOUGH: SEAS AT ONE HUNDRED FIFTY

Give me a lever long enough and somewhere to stand and I will move the world.

———————

—ARCHIMEDES (287–212 B.C.)[1]

I will believe that Columbia is over its science-averseness only when Archimedes is added to the list of classical worthies that scroll across the upper facade of Butler Library.

———————

—SALVATORE STOLFO, PROFESSOR OF COMPUTER SCIENCE, 2012[2]

F OR MUCH OF the twentieth century students and faculty alike came to Columbia's engineering school because it was part of a world-renowned university. This has been especially true for foreigners. The appeal for undergraduates has been reinforced by the presence of a distinctive core curriculum, which has assured them access to instruction of a high order not only in engineering, science, and mathematics, but in the humanities and social sciences, subject areas where the Columbia faculty has long excelled. Engineering alumni almost always cite "The Core" among their Columbia experiences they most cherish. Similarly, the school's location in New York City has served to attract students and faculty who might not have come but for its uniquely cosmopolitan setting.[3]

Thus, it is at least arguable that Columbia's engineering school has been for much of its history more the beneficiary of its positive institutional and locational circumstances than a primary contributor to them. To the extent that this assessment once held, it no longer does. Starting in the mid-1970s, the engineering school has not only benefited from Columbia's and

New York's improving circumstances but has contributed to those improvements. And now, on the eve of the engineering school's sesquicentennial, it has positioned itself to become a still more significant contributor to the university and the city, even as its faculty, students, and alumni stand ready to help move the world.

THE SEAS FACULTY AT ONE HUNDRED FIFTY

In 1939, the seventy-fifth anniversary of SEAS, the faculty consisted of thirty-eight men, born either in the United States, most in the northeast and most of them New York City, or Europe. All were Caucasian. Most attended Columbia as students. None had taught elsewhere, although several worked in industry before joining the SEAS faculty. Some continued to do so. Once joined, all but one stayed for the rest of their working days.[4]

SEAS faculty in 1939 defined themselves as engineers and educators. Some consulted with local industries and engineering firms. Few pursued research beyond what was required to secure their terminal degrees. The labs they and their graduate students operated on campus provided basic testing services to industry customers; they were not centers of innovation and invention or a source of federal revenue. Electrical Engineering Professor Edwin H. Armstrong and Chemical Engineering Professor Colin Fink were exceptional in their consuming research interests. Professor of Mechanical Engineering Charles Lucke, with his consulting practice downtown, and Civil Engineering Professor James Kip Finch, with administrative duties and historical studies, were more the norm. Neither they, nor the other native New Yorkers, nor the handful of European émigrés, had any professional incentive or cultural inclination to be anywhere else. Columbia was home.[5]

In 2013, on the eve of the SEAS one hundred fiftieth birthday, the faculty of 183 members included twenty-seven women (14 percent). A third were non-Caucasian, including three African Americans. Just over one-third were born in the United States (36 percent); another third in Asia and the Middle East; the rest in Europe or other non-U.S. locales. Most received their undergraduate training outside the United States, with only a handful of the American baccalaureates from Columbia. Most held American PhDs, with a third of these from California schools or MIT, while Columbia again accounted for a handful.[6]

SEAS faculty in 2013 see themselves as applied scientists. Careers center on their professional standing within subfields, which turn on peer evaluation of their research productivity and the promise of their graduate students. Graduate teaching is closely tied to their research, while undergraduate teaching, even if personally rewarding, is what they do to do their research. Most maintain active labs staffed by graduate assistants and research associates and funded by federal agencies that dispense grants through a competitive peer-reviewed process.[7]

Most SEAS faculty in 2013 first stepped foot on the Columbia campus when they came for their job interviews. For senior hires, Columbia was one step on a many-runged professional ladder. For junior hires, remaining at Columbia requires tenure, with the likelihood of securing it about one in two. Once tenured, staying at Columbia turns less on institutional loyalty and local ties than on the calculation of career prospects here against elsewhere. For today's SEAS faculty, Columbia is a contingent professional perch.[8]

Such a fundamental transformation occurred gradually. Armstrong and Fink were outliers in the 1930s as research-focused faculty, while subsequent postwar faculties included more of this type. Perhaps a quarter of the 1955 SEAS faculty identified themselves as applied scientists teaching at Columbia rather than as Columbia engineers. Half of these confirmed their contingent status by subsequently moving elsewhere.[9]

The crucial turning point occurred in 1979 with the creation of two departments: computer science, and applied physics and nuclear engineering (later APAM). Through subsequent appointments, these departments accelerated the social transformation of the SEAS faculty from mostly home-grown engineering teachers to a majority of foreign-born applied scientists, for whom research activity was paramount. Computer science would also lead the way in appointing and promoting women, who were themselves transformative agents.[10]

Helping this transformation were several developments, some national in scope, others particular to Columbia.

Among those occurring in the 1950s:

- the rapid postwar expansion of American higher education and the accompanying increase in engineering schools that would lay claim to national standing as producers of PhDs and active research centers;
- the rise in research funding from the federal government, competitively awarded;
- the adoption at Columbia of an "up-or-out" tenure system for faculty junior appointments;
- the elimination of extended faculty apprenticeships in which graduate students served as instructors, creating a local pool from which departments hired assistant professors;
- upping the requisite credential for appointment to engineering faculty from an MS or a professional degree to a DES or PhD.

Developments of the 1960s included:

- accreditation reviews that took a critical view of SEAS departmental "inbreeding";
- the development of statistically rigorous, national departmental rankings, starting with Allan Cartter's *An Assessment of Quality in Graduate Education* (1966).

In the 1970s:

- the adoption at Columbia of a university-wide ad hoc tenure review process that privileged outside evaluations of a candidate's national standing and publications, and where the final decision was no longer in the hands of the SEAS faculty or its dean but rested with the university provost;
- increased openness by SEAS to hiring foreign nationals with no prior exposure to Columbia, in response to a perceived decline in Americans pursuing advanced degrees and careers in engineering;

- the government-mandated opening of faculty hiring to women, whose credentials would be more universally assessed, along with the public posting of faculty openings as opposed to the earlier "old boys/hire my student" network;
- the rise and ubiquity of science citation indexes, which allowed comparative assessments of both faculty research activity and research impact;
- the stiffening of tenure requirements by Columbia's central administration, in part to contain the number of tenured faculty for budgetary reasons.

The 1980s:

- coupling SEAS faculty access to PhD students to their having ongoing research funding to support them;
- Columbia's aggressive efforts to capitalize on the patent licensing provisions of the Bayh-Dole Act, thus privileging by university officials of faculty entrepreneurial undertakings;
- the end of the Cold War, which led to a shift in federal research funding away from the physical sciences to the biological sciences.

The 1990s:

- arrival of the Internet, allowing instantaneous communication among research collaborators worldwide and the corresponding lessening of reliance on local contacts.[11]

All these developments produced a transnational SEAS faculty of research-focused applied scientists, in place of the earlier all-male teaching engineers. The operative question then becomes what brings these men and women to Columbia in the first place and what, besides their tenure, keeps them here? Extended conversations with SEAS faculty produced the following:

- an institutional environment conducive to research and stimulating teaching;

- likeminded colleagues of high quality and in sufficient numbers to permit specialization, shifts in research agendas, and collaboration;
- administrative leadership committed to the school's intellectual well-being and one with its professional self-identity as a school of applied science;
- academically motivated and globally recruited students in manageable numbers;
- competitive compensation and benefits;
- an urban environment that supports diversity;
- public recognition of faculty excellence in research and teaching;
- credible evidence allowing faculty to see the school growing stronger and being more favorably recognized;
- capacity to attract junior faculty from top graduate programs and senior faculty of international standing;
- finally, a university environment that values technology, applied science, and engineering and extends to a transnational SEAS faculty full university citizenship.[12]

To the extent these conditions exist, or are being put in place, Columbia's future as a global center of applied research and engineering education seems ensured.

COMING ON, PULLING EVEN

Comparisons may be invidious, but also instructive. Comparing SEAS past and present has been a preoccupation of the last three chapters, which urged the cumulative conclusion that the twenty-first-century SEAS is stronger, livelier, and more competitive than at any other time since the end of the nineteenth century. The comparison of SEAS with its engineering-school peers takes us into realms where the annual findings of *U.S. News and World Report* (*USNWR*) have become our one-eyed king. Suffice it here to suggest that, within the otherwise remarkably stable institutional ranking structure as *USNWR* portrays the competitive world of engineering education, SEAS's moves from forty-seventh in 2001 to twenty-fifth in 2005 to sixteenth in 2012 and to fifteenth in 2013 singles it out as a school on the rise.[13]

Realizing two presidentially stated goals for the school—to become the top ranked of the Ivy engineering schools and one of the nation's top ten engineering schools—appears within grasp. Cornell is the principal competition among the Ivies, although Columbia's undergraduate admissions statistics of late rank it the more selective. SEAS becoming one of ten top-ranked engineering schools in the nation will be harder, given the persistent size differences between SEAS departments and those of higher ranked schools.[14]

Some SEAS departments may soon be ranked in the single digits, with important specialties within some departments already there. A case in point: computer science. The thirty-five-member department in 2007 was almost three times the size Dean Likins had set as its outer limit— "Thirteen, that's it"—back at its founding in 1979. A substantial portion of this growth occurred through senior appointments, five by way of the implosion of the nearby Bell/AT&T Labs. In addition to Aho (1997), who specializes in mitigating software defects, these included Julia Hirschberg (2003), whose research is in spoken language processing; Vladimir Vapnik (2003), who works in machine learning; Mihalis Yannakakis (2004), a computational theorist, who came after a year at Stanford; and Steven Bellovin (2001), whose specialty is cybersecurity.[15]

Bellovin was already a member of NAE (2001) upon returning to Columbia, where he had been an undergraduate, joining Traub (1986) and Aho (1999), the other NAE members in the department. Since then Vapnik, who works in machine learning, was inducted in 2006; Nayar, who works in computer vision and graphics, in 2008; Yannakakis, a theorist, in 2011. This made computer science in 2013 the largest department in SEAS (with forty-three members in 2013), and with six NAE members and counting, the most nationally recognized. Meanwhile, in the new millennium, computer science has been successful appointing twelve junior faculty who have since been tenured, and who now constitute the department's third generation.[16]

Still among the smaller computer science departments vying for top ten ranking, the department has achieved international visibility in at least three subfields: spoken language processing; computer vision and graphics; and cybersecurity. Making it into the top ten is, a recent chair believes, "very achievable." Other developments suggest as much. The appointment in 2012 of Michael Collins, formerly of MIT, as the Vikram S. Pandit Professor of Computer Science, both adds strength to the spoken language process- ing group and to the department's recent engagement with social media. According to *Google Scholar*, between 2008 and 2012 Mihalis Yannakakis was Columbia University's second most cited faculty member. Meanwhile, the new Institute for Data Science and Engineering (IDSE), a cross-university research enterprise directed by Kathleen McKeown and Professor of Civil Engineering Patricia Culligan, has among its named participants some twenty departmental colleagues, while holding out the heady prospect of providing the department with the new appointments, space, and research outcomes to allow it to take that next big step.[17]

If computer science is the most likely prospect of an SEAS department breaking into the top ten as per *USNWR*, peer rankings compiled by other sources suggest upward movement elsewhere. In 2012 the IEOR department was ranked third by *Quantnet* among masters programs in financial engi- neering. In rankings controlled for faculty size, SEAS departments fare bet- ter. A case on point is rankings compiled by *Academic Analytics* and based on an average "academic productivity index" of the faculty in a given department. In the 2007 rankings of 375 mechanical engineering departments, SEAS's ranked fifth, tied with Stanford's department and just behind MIT's. These results may not have been so much "surprising," as reported, as prophetic.[18]

LOCAL BRAGGING RIGHTS

The comparison of SEAS departments with their Columbia science counterparts merits attention here because of it having historically been made, consistently to the detriment of SEAS. Indeed, it has informed the presumptive view that Columbia's engineers and their research pursuits were of a lesser order than Columbia's "real" scientists and helped relegate engineers to the periphery of the university's identity at the outset of the twentieth century, while assuring the "real" scientists a favored place at what most Columbians and Columbia observers saw as the university's central core. However, this once-defining and invidious ordering no longer holds water.[19]

As recently as the early 1990s the six cognate Columbia science departments were in the aggregate much larger than the engineering school. They had half again as many faculty, more undergraduate enrollments, more masters students, and twice as many PhD students. In 2013 these six science departments have only a few more regular members than the SEAS faculty. Whereas the SEAS faculty has almost doubled in size since 1995, that of the Columbia science departments has increased only slightly. Enrollment numbers for 2012–2013 show that more Columbia students receive instruction from SEAS departments than from the cognate science departments, at both undergraduate and masters levels. In 1995 the science departments produced ninety-one PhDs, while SEAS produced forty-five; in 2013 the science departments produced about the same number (ninety), while SEAS doubled its output to ninety-one.[20]

To be sure, the traditional relegation of the engineering school did not turn on such quantitative measures as faculty size and enrollments. The Columbia scientists' privileged place was perceived to be a function of their higher qualitative standing in their fields than Columbia engineers in theirs. At the most rarified level, this meant that science faculty and graduates won Nobel Prizes while engineering faculty and graduates did not. In point of fact, three SEAS graduates have won Nobel Prizes, but no faculty member until 1998, when Horst Stormer (a joint hire between SEAS and the physics department) did so only weeks after his arrival.[21]

Next in importance for perpetuating local bragging rights has been a comparison of the incidence of faculty and graduates admitted to the National Academy of Sciences. Founded in 1863, the NAS from its outset viewed engineers and applied scientists with such suspicion that membership has never been a fair test of the comparative standing within Colum-

FIGURE 10.1 Gordana Vunjak-Novakovic, native of Serbia, came to Columbia in 2005 as professor of biomedical engineering; specializes in human stem cells and tissue engineering. First SEAS faculty woman elected to the National Academy of Engineering (2012).

Source: SEAS Communications.

bia's larger scientific community. That did not stop it from being invoked, showing how Columbia's scientists have enjoyed a substantial numerical edge over its engineers. Only with the establishment of the National Academy of Engineering in 1967 has there been a recognition of distinction for Columbia engineers and applied scientists comparable to that enjoyed for more than a century by Columbia's other scientists. In 2013, Columbia's cognate science departments had thirty NAS members in their departments, while the SEAS had nine NAE members. [22]

The Nobel Prize and election to NAS or NAE typically come to scientists well along in their careers, often in recognition of their life's work. They are a lagging indicator of a given institution's relative standing. A metric more indicative of the moment is the relative success that SEAS and Columbia faculty currently have securing external funding for their research. Columbia administrators track three historical trend lines relating to sponsored research down to the school and department level by fiscal year: the number and dollar amount of proposals submitted by faculty; the number and dollar amount of grants awarded; and departmental expenditures covered by sponsored research projects. Proposals and awards data have been made available for FYs 2009 to 2012, while annual expenditures chargeable to sponsored research projects extend back to FY 1991. All three point to the same conclusion: the research activity of the SEAS faculty since the turn of the millennium grew at a rate that substantially exceeds that of the university overall.[23]

The longer time series, the annual SEAS expenditures against sponsored research projects, reflects two markedly different moments in the recent funding history of SEAS. The earlier, encompassing most of the 1990s (and likely extending further back into the 1980s), reflects at best a flat picture of the school's research portfolio. Between 1991 and 1999 the amounts annually charged hovered around $22 million, with three successive years (1993–1995) recording declines of between 6 and 8 percent. Beginning in 1998 (earlier in terms of actual awards), an upward trend takes hold that is thereafter sustained through to the most recent year for which expenditure numbers are available. In 2011–2012 these expenditures reached nearly $140 million.[24]

The second time series, encompassing FYs 2009 to 2012, provides data on the number and dollar amounts of research proposals submitted by SEAS departments to funding agencies, and the number and amounts of research awards. Both the proposals and grant trend lines indicate that the rate of seeking and securing research funding is accelerating and that the number of faculty annually submitting proposals is growing, while the gap between SEAS and the Columbia science departments has closed.[25]

The frequency of recent SEAS grant submissions varies by department, with the most proposal-active engineering departments, biomedical engineering and earth and environmental engineering, submitting proposals and securing grants at six or seven times the rate of the least active department, industrial engineering. But aside from IE/OR, which, as indicated earlier, is exceptional in having so few potential sources of externally sponsored research and funds its research largely through tuition revenues, all other SEAS departments—including the "mature" departments of chemical, civil, and mechanical engineering—are very much in the hunt for both federal and nonfederal funds to underwrite faculty research and summer salaries, to provide fellowship support for their PhD students, and to help cover the indirect costs of running the school and university.[26]

The pressure on SEAS faculty members to secure funding for their research is constant; failure to do so in one year (as one given over to administrative or public service) makes securing funding the next more difficult. To go without support for more than one funding cycle is to risk falling out of the universe of fundable researchers permanently. Some of the pressure to secure grants comes from deans, some from department chairs, some from one's colleagues. Salvatore Stolfo on computer science: "We created a culture of internal competition for others to emulate." Another source of the pressure is personal. Professors of mechanical engineering Jeffrey Kysar and James Hone have had great success in securing support for their research on the properties of grapheme, "the strongest material in the world," but as Kysar said of faculty maintaining labs: "Each and every one of us are running what is essentially a small consulting firm, with proposals to write, payrolls to meet, summers to work through, engineers to hire and then to send them off after they graduate."[27]

However competitive and personally trying the sponsored-research process, Columbia's engineers have become adept at it. They know and accept that the grant support they receive in one round requires them to produce tangible research outcomes to get favorable consideration in the next round of what, if sometimes viewed as a game professors play, is one without end—or at least coterminous with their lives as researchers.[28]

THE ENTREPRENEURIAL TURN

The real world today is located somewhere between the business school and the engineering school.

—IEOR PROFESSOR DAVID YAO, 2012

The occupational image of the American engineer into the 1970s was conservative and corporate. Successful careers came about by continuous upward movement through the ranks of a large firm, an Exxon or IBM or U.S. Steel, and assumed employment within the occupational sector, construction or pharmaceuticals or manufacturing, where that first job was secured. Risk taking and entrepreneurial instincts were optional, even suspect.[29]

The career paths of the SEAS graduates today are more likely to involve changing jobs and employers and occupational sectors several times. So by necessity and choice, young engineers have acquired a tolerance for new situations and a willingness to strike out on their own, often in directions far removed from what were once considered occupational circumstances appropriate to trained engineers. And if the entrepreneurial gene was largely missing from the traditional engineer's DNA, the genetic makeup of today's SEAS students and young graduates has been altered.[30]

Evidence for this sea change among faculty is provided by a survey conducted by Columbia's Office of Research Initiatives for Science and Engineering (RISE) in 2012 to gauge entrepreneurial activity across Columbia schools. The response rate among SEAS faculty (42 percent) was the highest among Columbia Morningside faculties, prompting RISE director Victoria Hamilton to conclude that "SEAS faculty were way more entrepreneurially disposed than they are given credit for." Their responses support that judgment:

—52 percent (of SEAS faculty) had been involved in starting or helping with an early stage company or organization;

—37 percent had been involved in starting a technology company;

—13 percent had been involved in starting a biotech company;

—13 percent had been involved in starting another type of company;

—24 percent had been company founders;

—15 percent had each raised more than $10 million for companies they've been involved with;

—87 percent know others who have been involved in starting technology-based companies.[31]

The entrepreneurial ethos may have penetrated some parts of the school more than others. Faculty in electrical engineering and computer science are generally cited as prime examples of professors with portfolios. They have more recently been joined by colleagues in biomedical engineering and industrial engineering. Nor is such activity limited to faculty. Graduate students also engage in entrepreneurship, sometimes in partnership with faculty and sometimes on their own, and even undergraduates come together to share ideas for bringing new products, processes, and services to market. Some of these gatherings are sponsored by the Columbia Engineering Entrepreneurship Mentoring Program and the Entrepreneurship Advisory Board, which "matches Columbia Engineering-educated business leaders with young entrepreneurs from across the University." Some of this activity is prompted by the ideals of social engagement and community building, some by the prospect of monetary reward, often by both.[32]

Another indicator of the strength of the entrepreneurial ethos at SEAS is its explicit inclusion in the curriculum. In 2010 the school authorized the IEOR department to offer an undergraduate minor in "Entrepreneurship and Innovation." Courses with "entrepreneurship" in their titles might be expected among those offered by the industrial engineering and operations research department, given its ties with the business school for three decades, but when departments once home to those who decried Michael Crow's solicitations as "the start of the baloney" now offer "Global Entrepreneurship in Civil Engineering," the theme can be said to have penetrated to the school's core.[33]

"ONLY CONNECT"

—E. M. Forster, Howards End[34]

Columbia's three most recent presidents have all emphasized that the university's future success in advancing research and providing effective instruction at the undergraduate, graduate, and professional levels will turn on the ability to overcome departmental and school boundaries by forging interdisciplinary and collaborative working relationships across campus(es). Departments and schools, faculty and students already forging those relationships represent Columbia's best hope for a future of continued productivity and service. And while it could not always be said of the engineering school, which for much of the twentieth century kept mostly to itself, SEAS arguably leads Columbia schools in acting on Forster's admonition, "only connect."

Part of what makes SEAS open to curricular collaborations is that, of Columbia's fifteen schools, it alone provides instruction to undergraduates, graduate students at both the MS and PhD levels, and to professional students. The computer science department teaches students in the College, Barnard, GS, Arts and Sciences, Journalism, and Law. Meanwhile, SEAS undergraduates take courses during their first two years in the Core from Columbia College faculty and their science coursework from faculty of arts and sciences. During their junior and senior years many SEAS students take elective courses at Barnard and the business school.[35]

Engineering faculty are natural collaborators in any undertaking that involves technology, and not much doesn't these days. Prime examples are the interest shown by the digital humanities faculty of the Columbia English department in the "big data" initiative emanating from the computer science department, the Journalism School's activities in digital communications, and the Business School's involvement with the MS program in Financial Management Systems run by the industrial engineering and operations research department.[36]

These interrelations between SEAS and the rest of the university can be envisioned as a series of overlaid network grids. The first overlay shows links between SEAS departments that are the product of joint appointments. In 2013, twenty-seven faculty from all nine departments held joint appointments in other engineering departments, with applied physics applied math and electrical engineering accounting for more than a third of them.[37]

A second overlay reflects faculty linkages that result from joint appointments between SEAS departments and Columbia science departments. In 2013, seven engineering faculty from five departments held joint appointments in one of four Columbia science departments, while three Columbia scientists were members of two engineering departments.[38]

A third overlay shows faculty linkages that result from joint appointments between SEAS departments and Columbia professional schools. Eleven SEAS faculty from four SEAS departments hold joint appointments in four different departments in health sciences and in the business school. Here the biomedical engineering department, with five faculty holding appointments in the radiology department of health sciences, leads the way, followed by industrial engineering, with its faculty links to the business school.[39]

A fourth overlay shows the linkages brought about by arrangements between SEAS faculty and four major trans-Columbia research centers, where they hold memberships and provide administrative leadership. These include the Earth Institute, with its earth and environmental engineering and civil engineering participants, the Center for Applied Probability, with its IEOR members, and a new research entity organized by returning Professor of Chemical Engineering Venkat Venkatasubramanian, the Center for the Management of Systemic Risk and Science (CMSRS).[40]

A fifth and last overlay reflects the linkages made through SEAS faculty affiliations with regional labs, among them the Brookhaven National Research Laboratory on eastern Long Island, the IBM Research facility in Westchester County, and the Bell (now Lucent) labs in New Jersey. While linkages with the latter two labs are relatively transitory, those with

Brookhaven, in the instances of the chemical engineer Jingguang Chen and the materials science professor Simon J. L. Billinge, constitute joint appointments.[41]

In all, almost a third of the 180 faculty in 2013 have an affiliation with a department, school, or research center that takes them outside their SEAS department in the conduct of their research and teaching. This proportion is especially striking when assistant professors, who make up about one-sixth of the faculty, are generally restricted to a single department affiliation so as not to unduly complicate the tenure process. While some of these extradepartmental links are in the nature of "courtesy appointments," and others may not last much beyond a given multiyear grant, all speak to the range of faculty interests that now transcend the traditional departmental boundaries. The total visual effect of these linkages is the very antithesis of the vertical structures of old, either of the silolike "little principalities and powers" that president Butler and Dean Barker lamented in the interwar years or the "steeple of excellence" that was civil engineering in the postwar era. They take the shape of a horizontal web, spreading itself in all directions so as to engage new research and instructional frontiers as they reveal themselves. An emergent case in point, neural engineering, a new field of research focusing on the workings of the human brain, currently engages SEAS faculty from biomedical engineering, electrical engineering, and applied physics and applied mathematics. Welcome to the future.[42]

Another function of these linkages, particularly those between SEAS departments, Columbia departments, and other professional schools, is that they provide the human tendons that hold Columbia together. This said, can there be, in the instance of transdepartmental collaboration, too much of a good thing? The SEAS department of earth and environmental engineering provides a possible instance.

The revival of the environmental movement in the 1960s made clear the need for environmental scientists and environmental engineers who would focus on resource sustainability, not simply perfecting extractive processes. Less clear was the organizational locus for producing them. The study of geology commenced at Columbia in 1865 in the School of Mines and the College, with the appointment of John Newberry as professor of geology and paleontology. Upon his retirement in 1892 geology migrated to the Faculty of Pure Science, with the engineering school retaining responsibility for mining, metallurgy, and mineral engineering. There matters stood until the founding of the Lamont Earth Observatory in 1947, under the energetic leadership of Maurice "Doc" Ewing. Soon thereafter, Lamont became the tail that wagged the geology-department dog. This asymmetry became even more pronounced in the 1960s when massive federal funding for ocean-floor exploration and an additional name-altering Doherty Foundation bequest of $7 million made the renamed Lamont-Doherty Earth Observatory Columbia's most prestigious research center. LDEO's physical remove, across the Hudson in Palisades, New York, only added to its stature and autonomy. Even after President McGill reasserted the university's authority over its endowment and effected Ewing's resignation in 1972, Lamont continued to overshadow the Morningside-based geology department.[43]

In 1996, at the direction of President George Rupp, the Columbia trustees created the Earth Institute, and appointed the economist Jeffrey D. Sachs its second director, with administration of the Lamont-Doherty Earth Observatory among his responsibilities. The relationship of the Earth Institute, located in Low Library, the department of geology in Schermerhorn, and the SEAS-based department of mining, metallurgy and mineral engineering in Mudd had yet to be determined. The assumption was, however, that their shared focus on earth science would encourage collaboration in research and instruction, with faculty and students utilizing the resources of all these entities to advance the stated goal of the Earth Institute to "address some of the world's most difficult problems, from climate change and environmental degradation, to poverty, disease and the sustainable use of resources."[44]

Collaboration proceeds apace. The 2013 staffing configuration of the Earth and Environmental Engineering Department attests to its entanglement with the Earth Institute and Lamont, and with a still more recent addition to the organizational mix, the Lenfest Center for Sustainable Energy. Of the EEE department's ten members, Peter Schlosser is codirector of the Earth Institute, Upmanu Lall is director of the Water Institute, Klaus Lackner is codirector of the Lenfest Center, and Ah-Hyung (Alissa)

Park is Lenfest Junior Professor in Climate Science. Minding the department are six faculty, one a newly appointed professor of practice who does much of the department's undergraduate teaching.[45]

Meanwhile, the absence of a departmental consensus on a suitable candidate for the Krumb Chair of Mining Engineering has left the chair unfilled for twenty-five years. Yet the department has been one of SEAS's most active and successful in securing grants and enjoys a healthy ratio of undergraduate majors per faculty member. Time will tell whether the earth and environmental engineering department will confirm the general proposition that SEAS departments are internally strengthened by the links they forge with other parts of the university—or be its limiting case.

THE SEAS STUDENT ESTATE AT ONE HUNDRED FIFTY

When I was an SEAS undergraduate, most of the classmates I knew were unhappy with the School. Not so today.

—AN SEAS PROFESSOR (ME 1986, PHD 1991) IN 2012 INTERVIEW[46]

I would not get in today.

—CHESTER LEE (SEAS CHEME 1970), FOUNDER OF
THE COLUMBIAN ASIAN ALUMNI ASSOCIATION, APRIL 2013[47]

In 1939 the SEAS undergraduate student body consisted of three classes of males, 195 in all, with another 100 male graduate students, half part-time. Nearly all came from within fifty miles of campus, from families of second-generation white ethnic New Yorkers, Italians, Irish, Jews, having graduated from one of the city's better public or parochial high schools. Most applicants were accepted and most accepted applicants enrolled. Students from Asian, African American, or Latino families were rare, as were foreign students. A few undergraduates came from families of means and lived on campus; most commuted, lived at home, and had jobs in their neighborhoods. Some were attracted to Columbia by the 3/3 plan that led to an AB and a professional degree after six years, but more opted for the quicker BS, or professional degree options that could be completed in four or five years. In 1939 SEAS awarded 106 degrees, 72 to graduating seniors and 34 MSs.[48]

In 2013 the SEAS student body consisted of 1,500 undergraduates, of whom 560 were women (36 percent), and 2,500 graduate students (27 percent women), most of them part-time. While a third of the undergraduates came from the New York metropolitan region, another third came from outside the northeast, with the remaining third foreign nationals. A substantial proportion attended private schools. Two-thirds of the graduate students were nonresident aliens, principally from India and China. Almost half of the U.S. residents were Asian American, while another 15 percent were African American or of Latino heritage. White American males made up about a quarter of the student body; white American females a sixth. Most undergraduates lived on campus; while about half received some form of financial aid. Most graduated in the prescribed four years. In 2013 SEAS awarded nearly fifteen hundred degrees: four hundred BSs, one thousand MSs, and ninety-one PhDs.[49]

Only 9 percent of the more than six thousand undergraduate applicants were admitted, with more than half (57 percent) of those admitted enrolling. This compares with a 31 percent acceptance rate in 2002, and a yield of 53 percent. The combined SAT scores of applicants accepted to SEAS increased from 2002, when they were below those accepted to Columbia College, to 2012, when they exceeded those of the College.[50]

The differences between SEAS students of 1939 and those of 2013 extend beyond the quantifiable. Most students in 1939 looked forward to careers as engineers in industry, while in 2013 many have their sights set on careers in law, medicine, or finance. Those heading to careers in business are already thinking about founding or joining start-up companies in the service sector rather than working for large corporations or engineering firms. Few students in 2013 secure professional licenses in the course of their education, while many engage in entrepreneurial ventures designed to jump-start careers of their own devising. Others envision themselves using their education to address global issues of sustainability with an idealism more reminiscent of the foreign-missionary-bound undergraduates of the nineteenth century than earlier and purportedly risk-averse engineering students.[51]

SEAS students today hold a different place on campus. Here the difference may not be as sharp when compared with their 1939 counterparts, most of whom began their Columbia careers in the College and were integrated into the undergraduate culture before entering the engineering school, but is dramatic when compared with students after 1959 entering directly into the engineering school. Those students had little contact with the College other than through the core curriculum, parts of which were then segregated by school. Some engineering students played intercollegiate sports alongside College teammates, but for the most part the two student bodies went their separate ways, with Columbia College men dominating the campus undergraduate scene.[52]

Several changes since have resulted in more equitable interschool social relationships among undergraduates. The easing of cross-registration restrictions in the 1970s led to greater classroom contact. The opening of Columbia College to women in 1983 brought women on campus, who crossed school lines to make friends and organize around gender-specific issues. Growing proportions of students in all four undergraduate programs identified themselves as minorities with organizational interests that focused on ethnic or racial identity irrespective of school affiliation. More recently, the narrowing gap between the selectivity of SEAS and the College to a point where both schools admit less than 10 percent of applicants—and the more recent gap in SAT scores favoring SEAS applicants—puts to rest any notion that getting into the College was harder than getting into SEAS. Engineering students continue to believe they work harder on their studies than their counterparts in the College or Barnard, but regard this widely held perception as a source of pride rather than a cause for grumbling. Finally, SEAS students today know their prospects of getting into law or medical or business school are as good as those of students from the College and Barnard, while the technical skills they take away at graduation are more likely to land them a job. Life is good.[53]

SEAS ALUMNI TODAY: BY THEIR FRUITS

By the spring of 2013, the living alumni of SEAS numbered nearly thirty thousand men and women. They play several important roles in the life of the school and university, although one less than in the past. Into the 1950s Columbia graduates made up 40 percent of the faculty; in 2013 only 5 percent of the faculty had undergraduate degrees from Columbia and those with graduate degrees, under 10 percent.[54]

As of 2013, twenty-four graduates of the engineering school have served as university trustees, with two, William Barclay Parsons and Frederick Coykendall, serving as board chairmen. Graduates of the engineering school (one least one at a time) have served continuously on the board since 1877, when F. Augustus Schermerhorn commenced his thirty-one-year tenure. Until 1909 Columbia trustees were elected to lifetime terms by the sitting trustees; after 1909, eighteen continued to be elected to life terms, with the other six elected to six-year terms as alumni trustees. Three engineering alumni so elected were subsequently elected to life terms.[55]

In addition to meeting their fiduciary responsibilities, several engineer trustees and their spouses have been among the school's most generous benefactors, including Henry Krumb, Edmund A. Prentis, and Armen A. Avanessians. Joining them as major benefactors of SEAS are alums Percy K. Hudson (EM 1899), Robert A. W. Carleton (CE 1904), Harvey Seeley Mudd (EM 1912), Seeley Greenleaf Mudd (BS 1919), and Morris A. Schapiro (EE 1925).[56]

Other engineer trustees have served their school through long and active involvement in the Columbia Engineering Alumni Association, which dates back to 1882. Anna K. Longobardo, one of the class of two in the first joint Barnard-Columbia Pre-Engineering Program, is a recent instance of an alumna active in the association's affairs who went on, in addition to a distinguished career as an engineer and executive of Unisys Corporation, to serve on the Columbia board. Other alumni have productively served on the Engineering Council, created in 1955 by the university trustees to "advise and assist the Trustees, the Faculty of Engineering, and the Dean in the development of the School," and on the Council's successor, the Board of Visitors, founded in 2007. All three organizations have in turn success-fully drawn on the school's global base of alumni for advice, guidance, and financial support.[57]

At different points in the history of SEAS individual alumni have come forward with specific recommendations for the school. Some have done so on their own initiative, as with F. Augustus Schermerhorn in 1882 and his "Committee of Ten." More often they have been called upon by the presi-dent, as in the instance of the Milton L. Cornell–led alumni committee in 1926; at other times it has been the dean who turned to them, as when Dean Galil in 2006 appointed a thirty-person alumni commission that pro-duced "A 2020 Vision for SEAS." President Butler regularly sought counsel from engineer-trustee Gano Dunn, president of the electrical manufacturer Crocker-Wheeler and later CEO of the construction giant J. G. White and Company, on matters affecting the university, as did President McGill the advice of engineer-trustee Robert D. Lilley (BS 1935), president of AT&T, in the wake of the 1968 disruptions.[58]

SEAS alumni also contribute to the school by the distinction of their careers. These include the three alumni, Irving Langmuir (MetE), in 1903; Robert C. Merton (BS 1965, OR), in 1997; and Alvin E. Roth (BS 1971, OR), in 2012, who have won the Nobel Prize. Two annual Columbia Alumni Asso-ciation awards honor these and other distinguished graduates. The Egleston Medal, first awarded on the occasion of the school's seventy-fifth anniver-sary in 1939, goes each year "to a graduate of the school who has significantly advanced his or her branch of the engineering profession, or management of engineering activities in general." Early recipients were often alumni on the engineering faculty who became leaders in their engineering specialties, such as Lotfi Zadeh (PhD 1952) in 2007 for "pioneering work in systems analysis, and subsequent development of fuzzy logic" and Masanobu Shinozuka (PhD 1959) as "Distinguished Professor, Chair of Civil and Environmental Engineering, University of California, Irvine; a dominant intellectual leader in establishing probabilistic mechanics, structural reliability, and risk assessment."[59]

Other Egleston medalists have included a father–son pair, Augustin L. Queneau (EM 1901), in 1960, a developer of recovery processes for rare metals, and his son Paul Queneau (BS 1933), in 1965, for discoveries in the field of metallurgy. In 1996, Anna K. Longobardo (BS 1948, MS 1951) received the Egleston medal for "setting standards of managing large complex global systems . . . , one of the first women to design, develop and evaluate control systems at sea on submarines and surface vessels. Trail blazer and innovator, both technically and managerially." Her husband, Guy Longobardo (BS 1947, MS, DES 1962), became the Egleston medalist a decade later, in 2006, for "advancing the discipline of bioengineering, especially in the area of respiratory disorders, applying the principles of applied mechanics and control theory to the field of physiology." Other recent recipients have included, in 2012, Bernard Roth (PhD 1963), cofounder of Stanford's Hasso Plattner Institute of Design and pioneering researcher in kinematics, robotics and design; and, in 2011, Michael Massimino (BS 1984), a NASA astronaut and developer of space shuttle robot arm operators.[60]

The second annual award, the Samuel Johnson Medal, was inaugurated in 2007 to honor an alumnus "for distinguished achievement in a field other than engineering." The first recipient was Ira M. Millstein (BS 1947, CU Law 1949), senior partner at the international law firm Weil, Gotshal & Manges and internationally recognized expert on corporate management; the 2012 Johnson medalist was Vikram Pandit (BS 1976, MS 1977, PhD 1986), university trustee and chief executive officer of Citigroup. Perhaps more than the Egleston Medal, the Samuel Johnson Medal speaks to the occupational diversity of younger SEAS alumni.[61]

By just about any indicator—alumni giving; attendance at reunion and alumni association events; interviewing prospective students; networking with, advising, and hiring graduates; participating in surveys; involvement in both school and university-sponsored special interest groups such as Society of Women Engineers and the Columbia University Asian-American Society—SEAS alumni are actively connected with SEAS and Columbia in proportions not seen since the early twentieth century. They join current students, faculty, and administrators in looking upon the school's future with reasoned confidence.[62]

THE MATTER OF DEANS

We don't need a good dean; we need a good algorithm.

—PROFESSOR OF COMPUTER SCIENCE, JONATHAN R. GROSS, 2012[63]

One of the central arguments of this book has been that for much of the history of Columbia's School of Engineering and Applied Science, strong administrative leadership has been wanting. This was due in part to structural factors, principally the limited authority vested in its dean by the University's central administration. But it has also been the result of factors internal to the School, among them entrenched faculty and occasionally alumni resistance to change. And in some instances, the problem has been personal in nature, a mismatch of a dean's strengths and the School's needs.

From the founding of the office in 1865, the SEAS dean has reported, at first, to the Columbia president and then later, through the university's chief academic officer, the provost. The first three deans had been members of the engineering faculty elected by their colleagues. Beginning in 1907, deans have been appointed by the president. Of the twelve subsequently appointed, four have come from the engineering faculty, two from the physics department, two from administration, and four from outside the university.[64]

SEAS deans have tended to proceed cautiously in dealing with senior faculty, especially department heads, whose chairmanships into the 1970s often extended over decades, and until then, negotiated their budgets and faculty appointments directly with the president, the trustee committee on education, or the provost. The dean's role in setting budgets and appointing faculty was advisory. When these structural inhibitions were compounded by questions of the dean's familiarity with the mission of the school, academic credentials, being "a real engineer," or with doubts about his commitment to the job, administrative drift ensued. Many of the structural limitations on the dean's authority were eliminated in the early 1970s by President McGill in his university-wide revamping of the budgetary arrangements. In turning to his professional school deans to find the revenues to stave off bankruptcy, he gave them budgetary and staffing powers unknown to previous deans. A reason McGill found Hennessy an unacceptable dean was that he was thought to be personally unwilling to use these new powers to effect the necessary results. In replacing him, McGill passed over two insiders, electrical engineer Ralph J. Schwarz and applied physicist Robert A. Gross. However devoted to the school and admired by his colleagues, Schwartz was thought to lack the personality to force change on fellow department chairs, while Gross, forceful and committed to change, lacked the support of the faculty. So McGill went outside and eventually settled on Peter Likins.[65]

The Likins deanship marked a decisive change in the history of SEAS. Before Likins, engineering deans had limited structural powers, sometimes further circumscribed by personal traits brought to the job, whereas Likins capitalized on the newly enhanced powers of the office and his forceful and persuasive personality to effect change and foster innovation. Likewise, Likins's successors Gross and Galil brought different personal strengths that contributed to the success of their deanships.

The Galil deanship benefited from an extended period of university-wide prosperity. Yet the unanimity of praise by current and retired faculty going back to the Dunning deanship suggest that personal factors figured in as well, not least his infectious enthusiasm for the challenge of securing SEAS a respected presence within the university. "Zvi gave us visibility" is a common theme in the reflections of both faculty and students who experienced his deanship. And with that visibility came enhanced standing.[66]

The case for the success of the Gross deanship is less certain in light of its downbeat ending, although his accomplishments have become widely acknowledged in retrospect. In the case of the Auston deanship, timing, personality, and faculty resistance conspired to leave much of his sweeping agenda to be implemented by his successor.[67]

This brings us to the recent deanship of Feniosky Peña-Mora (2009–2012). His personal story is compelling: A native of Santo Domingo, where he attended school and college, he learned English in his early twenties on summer visits to relatives in the upper Manhattan neighborhood of Washington Heights. Admitted to MIT as a graduate student, he earned a master's degree and a PhD in civil engineering in 1994. He then taught and held administrative positions at MIT for four years before moving to the University of Illinois in 2003, where he held a chaired professorship in the department of civil and environmental engineering and served as associate provost.[68]

The SEAS dean's search committee, created in the summer of 2007 and composed of alumni, students, and faculty, found much in Peña-Mora's career to commend him. In 1999 he won the National Science Foundation CAREER Award and the White House Presidential Early Career Award for Scientists and Engineers. In 2007, he won the Walter L. Huber Civil Engineering Research Prize of the American Society of Civil Engineers (ASCE). In 2008, he was recognized with the ASCE Computing in Civil Engineering Award for outstanding achievement and contribution in the use of computers in the practice of civil engineering.[69]

Peña-Mora's teaching and administrative experience at two of the nation's largest and highest ranked engineering schools was seen as another plus. The fact that he was born and raised outside the United States made his candidacy to lead a faculty the majority of which was foreign born and a student body more than half of which self-classified as racial minorities seemed timely. So did his Hispanic heritage in a city where Hispanics constituted the largest ethnic minority and Dominicans the fastest growing among them. Finally the obvious enthusiasm the forty-three-year-old candidate would bring to the job weighed in his favor. His name was included in the list of six finalists submitted to University Provost Alan Brinkley and President Lee Bollinger in the spring of 2009.[70]

In April 2009 President Bollinger announced the selection of Feniosky Peña-Mora as the fourteenth dean of the Fu Foundation School of Engineering and Applied Science. He was the third minority member appointed to a major Columbia administrative position that academic year, along with the African Americans Claude M. Steele as university provost and Michele M. Moody-Adams as dean of Columbia College. Prominent SEAS alumnus and search committee member Andy Gaspar (BS 1969) applauded the choice, describing Peña-Mora as "just what the School needed." Students were equally enthusiastic.[71]

Student opinion remained favorable fifteen months into his deanship, with a *Columbia Spectator* editorial crediting him with being "on the right track," while applauding his "energy, enthusiasm, and unconventionality." A picture of the dean charging down the aisle of 304 Havemeyer Hall to greet the incoming class of 2013 nicely captured the moment. A communications initiative designed to publicize the range and quality of the research of the faculty, *Excellentia, Eminentia, Effectio* (2011), attested to his ambitions for the school. Alumni seemed pleased with their new dean as evidenced by increases in annual giving from several benefactors who had been cultivated by earlier deans. In 2010, with Dean Peña-Mora at their side, SEAS alumnus and University Trustee Armen Avanessians and his wife Janette announced plans for a matching gift of $15 million to encourage alumni giving for endowed professorships.[72]

SEAS faculty proved a harder sell. Some department chairs immediately clashed with Peña-Mora over budgets, space and personnel, while others came to regard his strenuous efforts to boost the school's national visibility as costly and even unseemly. His announced goal of moving the school into the ranks of the top ten engineering schools in the country by 2020, although in line with the recent recommendations of "A 2020 Vision for SEAS," struck others as unrealistic and overly committed to a *U.S. News and World Report* view of the world.[73]

Two years into his deanship faculty distress with Peña-Mora went public with reports in *The New York Times* and *Columbia Spectator* that "Discord Over Dean Rocks Columbia Engineering School." A letter from senior faculty sent to Provost John H. Coatsworth in October 2011 and reprinted in the *Times* on December 7 declared the school's morale "is at an all-time low" and that only "a quick change in leadership" could repair the damage. Among the specific complaints: the dean's making financial considerations determinative in faculty appointments and promotions; hiring of space consultants without faculty involvement; the expansion of the MS program for budgetary reasons, doubling class size in some departments; the disavowal of oral agreements made with department chairs. "Not surprisingly, when one side routinely breaks agreements," the letter concluded, "the other side despairs of any future fruitful collaboration."[74]

Provost Coatsworth acknowledged the complaints but concluded that none reached the threshold of grounds for termination. Instead, he appointed Professor of Industrial Engineering Donald Goldfarb executive vice dean, to share administrative responsibilities with Dean Peña-Mora. Goldfarb, a member of the SEAS faculty for thirty-two years, had served as interim dean in 1994–1995 following Dean Auston's resignation. In the wake of the resignations of University Provost Claude Steele and Dean of Columbia College Michele Moody-Adams, some viewed the reluctance to fire Peña-Mora as not wishing him to become the third member of a racial minority to leave in the same year. Provost Coatsworth vigorously denied any such consideration. There the matter stood through early 2012, except for efforts on the Internet and in the press by members of the Hispanic and black communities of New York to defend Peña-Mora against charges they characterized as racially motivated.[75]

On July 3, 2012, Peña-Mora resigned his deanship, retaining his position as tenured member of the faculty, now as Edwin H. Armstrong Professor in the departments of civil engineering and computer science. "Differences of opinion are inevitable at times of change, and the criticism of a leader bringing about the change can be expected," the outgoing dean wrote, "particularly, I suppose, when the person is an outsider both institutionally and in other ways." Executive Vice Dean Donald Goldfarb became interim dean and the search for a new dean began.[76]

Though truncated, the Peña-Mora deanship encompassed three years (2009–2012) during which virtually every metric by which administrative success is measured registered gains. Research expenditures increased 36 percent, endowed professorships nearly doubled (from 23 percent to 45), the percentage of women faculty increased, as did that of non-Caucasians. On the student side of the ledger, undergraduate applications between April 2009 and April 2012 increased by 45 percent, while selectivity increased (from 14.4 percent acceptance to 9.4 percent) and yield went up (from 53 percent to 57 percent). Here, again, the percentage of women and underrepresented minorities in the entering classes increased, as did SAT scores by 21 points. Graduate applications nearly doubled, while the GRE scores of those admitted to both the MS and PhD programs increased. Many of these indicators of institutional well-being were on the rise when Peña-Mora assumed the deanship, allowing the possibility that their continued upward trending was the result of institutional momentum more than his administrative leadership. But to conclude would be to miss the principal achievement of the Peña-Mora deanship: it extended the upward swing of Columbia's engineering school into an unprecedented third decade.[77]

On March 26, 2013, President Lee Bollinger announced the appointment of Mary Cunningham Boyce as the fifteenth dean of SEAS, effective

FIGURE 10.2 Mary C. Boyce, appointed thirteenth dean of SEAS in March 2013 by Columbia University President Lee Bollinger (2004–). Had been MIT professor of mechanical engineering and 2012 inductee to National Academy of Engineering.

Source: SEAS Communications.

July 1. Boyce had been the Ford Professor of Mechanical Engineering at MIT, where she had taught for twenty-four years following completion of her PhD there in 1987. As SEAS's first female dean, Boyce extended the string of deans as "firsts" to five (Gross the first Jewish dean; Auston the first foreign-born dean; Galil the first dean born outside North America; Peña-Mora the first dean of Hispanic heritage). More tellingly, as chair of a top-ranked sixty-person department and a recently elected member of the National Academy of Engineering (2012), Boyce brought to her new position a combination of administrative experience and scholarly credentials unsurpassed by any of her predecessors. One thing for sure, the days when someone could say "No one at MIT takes the Columbia engineering school seriously. No one knows who the Dean is," had ended.[77]

"IN THE CITY OF NEW YORK"

The letter to Provost Coatsworth in October 2011 calling for firing of the dean acknowledged doing so might complicate the resolution of another controversial question that had Columbia and its engineering school in the news: How, or even whether, should Columbia respond to Mayor Michael Bloomberg's December 2010 call for proposals to compete for upwards of $100 million in city money and an estimated $300 million in city-owned property to establish "a world-class graduate school of technology and applied science in New York City"?[78]

Columbia seemed reluctant to take up the invitation, while Cornell, Stanford, Carnegie Mellon, NYU, and several foreign universities scrambled to present proposals, some free-standing and some partnering with other universities, foreign and city-based. By the July deadline for applying, the city's Economic Development Corporation acknowledged receipt of seven credible proposals. Among the frontrunners, Cornell, with its partner Technion-Israel Institute of Technology, proposed to invest $2 billion of their own money to build a 45-acre campus on Roosevelt Island; Stanford proposed a $1.25 billion investment in 1.9 million square feet of floor space in lower Manhattan. Columbia's proposal was among the last submitted and the most modest in scope. Also, whereas Stanford and Cornell both planned to build on the city-owned sites, Columbia proposed a data science institute on its own 18-acre Manhattanville site.[79]

Columbia's hesitation was not, as an editorial in the *Columbia Spectator* darkly suggested, "an unspoken concession . . . that SEAS is not a 'top-caliber' engineering school." That Cornell was headed by an engineer and Stanford by an MD-bioengineer, whereas the Columbia president was a constitutional scholar, may account for some of the difference in enthusiasm the three presidents brought to the competition. More important was President Bollinger's considered reluctance to set aside an expansion already under way for the Manhattanville campus to link Columbia's uptown health sciences schools with the science departments and engineering school on Morningside, in favor of a taking a flyer on further dispersing Columbia's sciences by vying for money to be spent on city land in lower Manhattan.[80]

Stanford's President John L. Hennessy, who had seemed ready to bet the Palo Alto "farm" on winning the competition, three days before the announcement of the winner was instructed by Stanford trustees to drop out of the contest. Cornell continued to see the competition as an opportunity to link its engineering school in out-of-the-way Ithaca with its health sciences already in the city. And unlike Stanford, Cornell had its alumni, many of whom lived and worked in New York, fully on board. One came up with gift of $350 million in the last week to sweeten Cornell's bid. On December 19, 2011, to no one's surprise, Cornell was declared the winner.[81]

In doing so, Mayor Bloomberg indicated that the city would consider funding parts of the submitted proposals from New York–based institutions at a future date. In April 2012, New York University announced its plan to build a Research Institute in Brooklyn, for which it was to receive $50 million from New York City. And then in August, the mayor announced a $15 million gift to Columbia to assist in the construction of the Columbia Institute for Data Sciences and Engineering on its Manhattanville campus. Although less than Cornell and NYU came away with, Columbia had the satisfaction of getting its money for a project it was already committed to, rather than to an open-ended project of uncertain length, cost, and utility.[82]

FIGURE 10.3 Donald Goldfarb, professor of industrial engineering; Kathleen McKeown, professor of computer science; Patricia Culligan, professor of civil engineering, interim dean (2012–2013); and Michael Bloomberg, New York City mayor (l to r), on 2013 occasion of a $15 million award to SEAS for its planned Institute of Scientific and Engineering Data (ISED), with McKeown and Culligan as founding codirectors.

Source: SEAS Communications.

Other consolations can be extracted from the NYC engineering competition. It gave Columbia yet another opportunity to vouchsafe its fealty to the city of New York and to Manhattan, if not downtown where it began. The publicity attending the competition focused attention of Columbians on their often overlooked engineers. *The Chronicle of Higher Education*, in handicapping the contenders, provided the conventional view: "[Columbia] is better known in New York for its business, law, and medical schools than for technology entrepreneurship." That's pretty much as the *Christian Examiner* sized it up 158 years earlier—"good in classics, weak in sciences." That perception is changing; the reality already has.[83]

One of the latent consequences of the Bloomberg competition has been to spark two related conversations: the place of Columbia in the city of New York, and the place of engineering within Columbia. Not since the 1880s, when New York became a world center of industry, construction, and transportation, have Columbia's engineers been as central to the city's aspirations to become a world center of technology and information, even as it fights to maintain its standing as the world center of finance and communication.[84]

Internally, the competition provided Columbia an opportunity to showcase its latest and most ambitious initiative in technology and Columbians a chance to consider its implications. Primarily the work of Kathleen McKeown of the computer science department, along with Patricia Culligan of the civil engineering department, the Institute for Data Sciences and Engineering (IDSE) envisions precisely the kinds of interdisciplinary and interschool collaboration that is likely to be the future of Columbia. As presently configured the institute consists of six centers—Health Analytics, Financial Analytics, Smart Cities, New Media, Cybersecurity, and Foundations of Data Sciences—each with a chairperson and a committee, plus affiliated members. With McKeown as director and Culligan as associate director, the number of Columbia faculty involved approaches 180, with some faculty involved in more than one center. Among Columbia professional schools represented are health sciences, business, law, public and international affairs, social work, architecture and historic preservation, and journalism; among departments in arts and sciences are biology, chemistry, economics, history, and statistics. Yet more striking than the sheer number of participants or the range of their school affiliations is the place and preponderance of SEAS faculty among them. Four of the six centers are chaired by SEAS faculty, while more than half the committee members and center affiliates are SEAS faculty, drawn from all nine departments.[85]

One could well end this history with the research implications and promise of the "big data" initiative. But just as fitting is to end with two curricular initiatives. The first is the growing practice of appointing faculty, usually as professors of practice exempted from the up-or-out requirements of a tenure-track position, to departments where their primary mandate is attending to the undergraduate curriculum. These positions, by 2013 adopted by all nine departments, when combined with the ongoing practice

FIGURE 10.4 Northwest Science Building, viewed from the southeast, opened on the Morningside campus in 2011, the location of research labs of several engineering faculty and the temporary home of the Institute of Scientific and Engineering Data.

Source: SEAS Communications.

of the careful monitoring of student evaluations by department chairs and the dean, ensure that undergraduate teaching remains a high priority in the SEAS of tomorrow.[86]

Another current curricular initiative comes in a sector in which Columbia's engineers have long been identified, that of distance learning. The

FIGURE 10.5 Michael Collins, professor of computer science; Garud Iyenfar, professor of industrial engineering; Yannis Tsividis, professor of electrical engineering; and Martin Haugh, codirector of the Center for Financial Engineering (l to r), all offering Massive Open Online Courses (MOOCs) in 2013.

Source: SEAS Communications.

Columbia Video Network, launched in 1986, has provided thousands of off-site students access to SEAS courses delivered first in video formats and more recently digitally over the Internet. One form of online instruction has recently attracted new attention, along with a new acronym, "MOOCS," for "massive open online courses." Whether, as some academic administrators

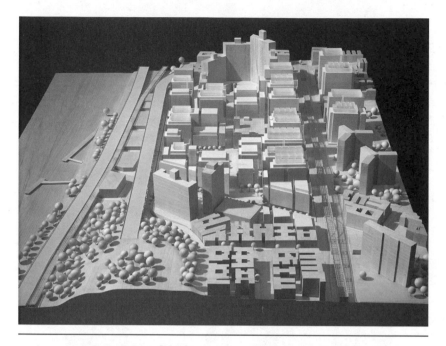

FIGURE 10.6 Manhattanville campus site north of the
Morningside campus being viewed by SEAS Dean Mary C. Boyce
from the Northwest Science Building. Future home of the
Institute of Scientific and Engineering Data (ISED).

Source: SEAS Communications.

warn darkly, MOOCS presage "the end of education as we know it," or as
The New York Times less apocalyptically calls them, "a grand experiment,"
they represent a technology-enabled curricular development that Columbia
and other universities must take seriously. Accordingly, SEAS introduced
Columbia's first three MOOCs in the spring of 2013, taught by four faculty
drawn from three departments, with more to come.

So let the university's then newly appointed and tech-savvy chief digital
officer, Sree Sreenivasan, who has since left, have the last word with a com-
ment he made specifically about the challenge of online learning at today's
Columbia, but applied here more generally: "The Engineering School made
perfect sense to start. . . . It's so far ahead of the rest of the campus."[87]

NOTES

PREFACE

1. Rosalind Williams, *Retooling: A Historian Confronts Technological Change* (Cambridge: MIT Press, 2002), 13.
2. C. P. Snow, "Two Cultures," *The Two Cultures and a Second Look* (Cambridge: Cambridge University Press, 1964.

I. ENGINEERING IN AMERICA–BEFORE ENGINEERS

1. As quoted in Peter L. Bernstein, *Wedding of the Waters; The Erie Canal and the Making of a Great Nation* (New York: Norton, 2001), 317.
2. "Recent Troubles at Columbia College," [Boston] *The Christian Examiner*, June 1854, as quoted in Milton Halsey Thomas, "The Gibbs Affair at Columbia in 1854" (MA thesis, Columbia University, 1942).
3. For present purposes, "engineering" is defined as a set of technical undertakings intended to effect constructive change in the standing physical order; an "engineer" is someone with the requisite training, occupational experience, and knowledge of the relevant science and mathematics to purposefully engage in these undertakings. What constitutes the requisite training, experience, and knowledge depends upon the period under consideration and the complexity of the undertaking/process. Well into the 19th century in America, major engineering projects were conducted by self-taught/learn-on-the-job "engineers," typically with no formal academic training; the careers of Thomas Edison and Henry Ford attest to the persistence of the autodidact engineer into the 20th century.

4. William Cronon, *Changes in the Land: Indians, Colonists, and the Ecology of New England* (Hill & Wang, 1983); W. Jeffrey Bolster, *The Mortal Sea: Fishing the Atlantic in the Age of Sail* (Cambridge: Harvard University Press, 2012).

5. Terry S. Reynolds, "The Engineer in 19th century America," in Terry S. Reynolds, ed., *The Engineer in America* (Chicago: University of Chicago Press, 1991), 7–26.

6. William Bradford, *Of Plymouth Plantation, 1620–1647*, ed. Samuel Eliot Morison (Modern Library, 1967); on Winthrop, Francis Bremer, *John Winthrop: America's Forgotten Founding Father* (New York: Oxford University Press, 2003); on Williams, Edwin Gaustad, *Roger Williams* (New York: Oxford University Press, 2005).

7. Joyce Chaplin, *The First Scientific American: Benjamin Franklin and the Pursuit of Genius* (Basic Books, 2006).

8. James Kip Finch, *Early Columbia Engineers: An Appreciation* (New York: Columbia University Press, 1929), 5–11.

9. F. Daniel Larkin, *John B. Jervis: An American Engineering Pioneer* (Ames: Iowa State University Press, 1990).

10. Ronald E. Shaw, *Canals for a Nation: The Canal Era in the United States, 1790–1830* (Lexington: University of Kentucky Press, 1990).

11. Bernstein, *Wedding of the Waters*, 24–48.

12. Brook Hindle, *Engines of Change: The American Industrial Revolution, 1790–1860* (Washington, D.C.: Smithsonian Institution, 1986), 59–73.

13. Arthur H. Clark, *The Clipper Ship Era* (New York: Putnam's, 1910); John H. Morrison, *History of American Steam Navigation* (New York; Stephen Days Press, 1958).

14. Robert G. Albion, *The Rise of New York Port, 1815–1860* (New York: Scribner's, 1939).

15. On Croton Aqueduct, Kenneth T. Jackson, ed., *The Encyclopedia of the City of New York* (New Haven: Yale University Press).

16. Ray Billington, *Westward Expansion* (New York: Macmillan, 1967), 574–634.

17. On Horatio Allen, see F. Daniel Larkin, *John B. Jervis: An American Engineering Pioneer* (Ames: Iowa State University Press, 1990).

18. James Kip Finch, *Early Columbia Engineers: An Appreciation* (New York: Columbia University Press, 1929), 27–32.

19. On Samuel F. B. Morse, Brook Hindle, *Emulation and Invention* (New York: Norton, 1983).

20. On Carnegie, David Nasaw, *Andrew Carnegie* (New York: Penguin, 2006); on Edison, Paul Israel, *Edison: A Life of Invention* (New York; Wiley, 1998).

21. Billington, *Westward Expansion*, 617–634.

22. Alfred D. Chandler, *The Visible Hand: The Managerial Revolution in American Business* (Cambridge: Harvard University Press, 1977); Bruce Sinclair, "Science, Technology, and the Franklin Institute," in Alexandra Oleson and Sanborn C. Brown, eds., *The Pursuit of Knowledge in the Early American Republic* (Baltimore: Johns Hopkins Press, 1976), 194–207.

23. John S. Whitehead, *The Separation of College and State: Columbia, Dartmouth, Harvard, and Yale, 1776–1876* (New Haven: Yale University Press, 1973).

24. Brook Hindle, *The Pursuit of Science in Revolutionary America, 1735–1789* (Chapel Hill: University of North Carolina Press, 1956), 84–85, 89, 91.

25. Robert A. McCaughey, *Stand, Columbia: A History of Columbia University in the City of New York, 1754–2004* (New York: Columbia University Press, 2003), 25–26; Joseph J. Ellis, *The New England Mind in Transition: Samuel Johnson of Connecticut, 1696–1772* (New Haven: Yale University Press, 1973; Norman S. Fiering, "President Samuel Johnson and the Circle of Knowledge," *William and Mary Quarterly* 28 (April 1971): 199–236.

26. Herbert Schneider and Carol Schneider, eds., *Samuel Johnson, President of King's College: His Career and Writings* (New York: Columbia University Press, 1929), IV, 222–224.

27. McCaughey, *Stand, Columbia*, 30, 45; David C. Humphrey, *From King's College to Columbia, 1746–1800* (New York: Columbia University Press, 1976), 106–108, 166–169, 206–207.

28. McCaughey, *Stand, Columbia*, 44–45.

29. On Stevens family and bequest, see Archibald Turnbull, *John Stevens: An American Record* (1928).

30. Hindle, *Pursuit of Science*, 84–86, 90.

31. Stanley M. Guralnick, *Science and the Ante-Bellum American College*, Memoir No. 109 (Philadelphia: American Philosophical Society, 1975).

32. McCaughey, *Stand, Columbia*, chaps. 2 and 3.

33. Edwin Layton, "Mirror-Image Twins: The Communities of Science and Technology in 19th-Century America," *Technology and Culture* 12, no. 4 (October 1971): 562–580.

34. D. Graham Burnett, *Trying Leviathan: The Nineteenth-Century New York Court Case That Put the Whale on Trial and Challenged the Order of Nature* (Princeton: Princeton University Press, 2007).

35. *The Diary of George Templeton Strong*, Allan Nevins and Milton Halsey Thomas, eds., 4 volumes (New York: Macmillan, 1952), 1: 3.

36. McCaughey, *Stand, Columbia*, 99–101.

37. Layton, "Mirror-Image Twins," 562–580.

38. James Kip Finch, *Early Columbia Engineers: An Appreciation* (New York: Columbia University Press, 1929), 17–22.

39. Strong, *Diary*, 1:2, 3, 5, 6, 7, 10, 15, 16, 22–24, 29, 37–40, 42, 43, 50, 51, 58, 59, 70, 77, 85, 87, 150, 153, 187, 296. On Peirce and Harvard of his day, Robert A. McCaughey, "The Transformation of American Academic Life: Harvard University, 1821–1892," *Perspectives in American History* 8 (1974): 239–332.

40. On the midcentury mobilization of the American scientific community, Sally Gregory Kohlstedt, *The Formation of the American Scientific Community: The American Association for the Advancement of Science* (Urbana: University of Illinois Press, 1976).

41. Strong, *Diary*, 2: 135–137, 141.

42. McCaughey, *Stand, Columbia*, 120–130.

43. Ibid., 120–130.

44. Ibid., 125.

45. "One of Its Trustees" [Samuel B. Ruggles], *The Duty of Columbia College to the Community: And Its Right to Exclude Unitarians from Its Professorships of Physical Science* (New York: John F. Trow, Printer, 1854), 22.

46. Strong, *Diary*, II, 172.

47. *Christian Examiner, 1854,* quoted in Thomas, "Columbia and the Gibbs Affair in 1854."

48. F. W. Clarke, "Biographical Memoir of Wolcott Gibbs, 1822–1908" (Washington, D.C.: National Academy of Sciences, 1910).

49. McCaughey, *Stand, Columbia,* 143.

50. Russell H. Chittenden, *History of the Sheffield Scientific School of Yale University, 1846–1902,* 2 vols. (New Haven: Yale University Press, 1928).

51. On the midcentury mobilization of the American scientific community, Kohlstedt, *The Formation of the American Scientific Community.*

52. Carl Diehl, *Americans and German Scholarship, 1770–1870* (New Haven: Yale University Press, 1978).

2. FAST START 1864–1889

1. *The Diary of George Templeton Strong,* Allan Nevins and Milton Halsey Thomas, eds., 4 vols. (New York: Macmillan, 1952), 4:529.

2. Frederick A. P. Barnard, March 2, 1874, Columbia Trustee Minutes, Vol. 7, Columbiana Archives, Rare Books and Manuscript Library, Columbia University.

3. Anonymous, "Review of The American University," *The School of Mines Quarterly* 5 (Fall 1884): 169–170.

4. Edna Yost, "Egleston, Thomas." *Dictionary of American Biography* (New York: Scribner's, 1928–1936), 12:356; (Egleston award winner) James Kip Finch, *A History of the School of Engineering* (New York: Columbia University Press, 1954); Egleston's papers are in the Columbia University Archives.

5. Ibid.

6. Strong, *Diary,* 3:316.

7. George Templeton Strong's *Diary* is the best source on the founding and early years of the School of Mines. Because James Kip Finch did not use it in compiling his *History of the School of Engineering* (1954, at the time of the university's bicentennial), it is quoted extensively here. The complete manuscript version of the diary is at the New York Historical Society.

8. Strong, *Diary,* 3:367; Columbia College Trustee Minutes, October 15, December 21, 1863; McCaughey, *Stand, Columbia,* 143.

9. Strong, *Diary,* 3:383; McCaughey, *Stand, Columbia,* 143.

10. William J. Chute, *Damn Yankee! The First Career of Frederick A. P. Barnard, Educator, Scientist, Idealist* (Port Washington, NY: National University Publishers, 1978); John Fulton, *Memoirs of Frederick A. P. Barnard, Tenth President of Columbia College in the City of New York* (New York: Macmillan, 1896); Stacilee Ford Hosford, "Frederick Augustus Porter Barnard: Reconsidering a Life" (EdD dissertation, Teachers College, Columbia University, 1991).

11. Minutes of the Columbia College Trustees, April 6, 1863, February 1, 1864; Strong, *Diary*, 3:398.

12. On Vinton appointment, Strong, *Diary*, 3:524; on Chandler appointment, Minutes of the Trustees of Columbia College, September 12, 1864.

13. Strong, *Diary*, 3:524, 543

14. Ibid., 4:516.

15. Minutes of the Trustees of Columbia College, September 14, 1864.

16. Finch, *School of Engineering*.

17. Strong, *Diary*, 4:436, 528–529.

18. Strong, *Diary*, 3:487.

19. Strong, June 9, 1864, 3:457. Strong on Barnard continued.

20. Strong, *Diary*, 3:510.

21. Trustee critics of School of Mines at outset in Treasurer Gouverneur Morris Ogden.

22. McCaughey, *Stand, Columbia*, 155–176.

23. Ralph Waldo Emerson, "Self-Reliance," *Emerson's Essays, First and Second Series*, ed. Irwin Edman (New York: Thomas Crowell, 1926).

24. Robert L. Larson, "Charles Frederick Chandler: His Life and Work," Columbia PhD dissertation, 1950; Marston Taylor Bogert, "Charles Frederick Chandler, 1836–1925," *Biographical Memoirs* 14 (Washington, D.C.: National Academy of Sciences, 1931), 125–181; Margaret Rossiter, "Charles F. Chandler Collection—Preliminary Inventory," Columbia University Libraries, Rare Book and Manuscript Library (1973).

25. Bogert, "Chandler."

26. Ibid.

27. Strong, *Diary*, 3:485.

28. Larson, "Chandler," 58.

29. Ibid.

30. Ibid.

31. Leader of American Chemical Society.

32. Bogert, "Chandler."

33. Ibid.

34. On Barnard as College skeptic, McCaughey, *Stand, Columbia*, 156; on Chandler, Larson, *Chandler*.

35. Larson, *Chandler*; McCaughey, *Stand, Columbia*, 152.

36. Strong, *Diary*, 4:516.

37. Minutes of the Trustees of Columbia College, March 2, 1874.

38. Strong, *Diary*, 4:528–529.

39. Minutes of the Trustees of Columbia College, May 4, 1877.

40. Minutes of the Trustees of Columbia College, March 4, 1878.

41. Minutes of the Trustees of Columbia College, May 6, 1878; June 6, 1881.

42. McCaughey, *Stand, Columbia*, 163–170; Rosalind Rosenberg, *Changing the Subject: How the Women of Columbia Shaped the Way We Think About Sex and Politics* (New York: Columbia University Press, 2004), 32–47.

43. Minutes of the Trustees of Columbia College, November 5, 1888.

44. "School of Mines Notes," *School of Mines Quarterly* 3 (1881–1882): 57.

45. On trustee rejection of proposed name change (F. A. P. Barnard), *President of Columbia College Annual Report, June 1881*, 59.

46. Minutes of the Trustees of Columbia College, February 3, 1883.

47. "Enrollments," *School of Mines Quarterly* 3 (1881–1882).

48. Minutes of the Trustees of Columbia College, May 4, 1877.

49. *The School of Mines Quarterly* 1 (1880): 223–225.

50. School of Mines Quarterly 1 (1879–1880), 223.

51. "Editors' Note," *School of Mines Quarterly* 1, no. 1 (November 1879).

52. Minutes of the Trustees of Columbia College, February 4, 1877; "School of Mines Notes," *The School of Mines Quarterly* 2, no. 4 (1880–1881): 223.

53. Minutes of the Trustees of Columbia College; on early American PhDs, see *A Century of Doctorates* (Washington, D.C.: National Academy of Sciences, 1978).

54. Minutes of the Trustees of Columbia College, June 2, 1873; June 7, 1875. For impression that School of Political Science inaugurated graduate degree programs at Columbia, see Gordon Hoxie, ed., *A History of the Faculty of Political Science, Columbia University* (New York: Columbia University Press, 1954).

55. Rosenberg, *Changing the Subject*, 45–46, 53–54, 325.

56. "Editors' Note," *The School of Mines Quarterly* 1 (November 1879–May 1880).

57. *The School of Mines Quarterly* 3 (1881–1882).

58. For degree statistics, see http://engineering.columbia.edu/leverlongenough.

59. For the occupational outcomes of 1867–1881 graduates, see *The School of Mines Quarterly* 3 (1881–1882): 242; Samuel B. Christy, "The Growth of American Mining Schools and Their Relation to the Mining Industry," *Transactions of the American Institute of Mining Engineers* (1893); see also the book of one of the first graduates of the School of Mines, John A. Church (EM 1868), *Mining Schools of the United States* (New York, 1871).

60. For the geographic dispersion of graduates in 1882, *The School of Mines Quarterly* 4 (1882–1883): 70–72.

61. School of Mines Quarterly 3 (1881–1882), 242.

62. http://engineering.columbia.edu/leverlongenough

63. On Hollerith, Geoffrey D. Austrian, *Herman Hollerith: Forgotten Giant of Information Processing* (1982).

64. Mines graduates who became Columbia professors: Henry S. Munroe, 1869; John K. Rees, 1875; Frederick R. Hutton, 1876; Francis B. Crocker, 1882; Alfred J. Moses, 1882; Robert Peele, 1883; Michael I. Pupin, 1883; Arthur L. Walker, 1883; James F. Kemp, 1884; Ira Woolson, 1886; Lea M. Luquer, 1887; Robert M. Raymond, 1889.

65. "Committee of Ten Report," *The School of Mines Quarterly* 5 (1884–1885): 177–184.

66. John W. Burgess, *Reminiscences of an American Scholar* (New York: Columbia University Press, 1934; 1966); Hoxie, ed., *Faculty of Political Science, Columbia* (New York: Columbia University Press, 1954).

67. Burgess, *Reminiscences*, 242.

68. Burgess, *The American University: When It Shall Be? Where Shall It Be? What Shall It Be?* (New York: Ginn, Heath, 1884) (reprinted in Burgess, *Reminiscences*, 349–368).

69. Burgess, *The American University*, 362–364

70. Ibid., 366.

71. Thomas H. Huxley, "Science and Culture" (1880), 26, as quoted in Paul Forman, "The Primacy of Science in Modernity, of Technology in Postmodernity, and of Ideology in the History of Technology," *History and Technology*, 23, no. 1 and 2 (2007): 1–152. See also Ronald Kline, "Construing 'Technology' as 'Applied Science': Public Rhetoric of Scientists and Engineers in the United States, 1880–1945," *Isis* 86, no. 2 (June 1995): 194–221.

72. Henry Rowland, "The Highest Aims of the Physicist," *Science* 10 (1899): 825–833.

73. "Review of John W. Burgess, The American University," *The School of Mines Quarterly* 6 (1884–1885): 169–170.

74. Ibid., 171.

3. A CORNER IN THE UNIVERSITY 1889–1929

1. Seth Low, *President's Annual Report*, October 1891, CUA.

2. Joseph Barker Papers, Box 359, CUA.

3. Edwin E. Slosson, *Great American Universities* (New York, 1910); McCaughey, *Stand, Columbia*, chap. 7.

4. Seth Low, *President's Annual Report*, October 6, 1890, CUA.

5. Nicholas Murray Butler, interview with Dwight Miner, December 3, 1940, Dwight Miner Papers, Box 4, RBML, CU.

6. Gerald Kurland, *Seth Low: The Reformer in an Urban and Industrial Age* (Boston: Twayne, 1971); on Low's civic career, David C. Hammack, *Power and Society: Greater New York at the Turn of the Century* (New York: Columbia University Press, 1982).

7. Kurland, *Seth Low*.

8. McCaughey, *Stand, Columbia*, 180–182.

9. Ibid., 182–187.

10. Ibid., 187–192.

11. John Henry Van Amringe, "The School of Mines," *The School of Mines Quarterly* 10 (1888–1889): 338–350.

12. On proposed names for engineering school, Trustee Minutes, February 2, 1891, CUA.

13. Trustee Minutes, March 8, 1891.

14. On the Da Costa bequest, Trustees Minutes, March 18, 1889.

15. Trustee Minutes, March 2, 1891.

16. Eric Kandel, "Biology at Columbia," and Qais Al-Awqati, "Edmund Beecher Wilson: America's First Cell Biologist," in Wm. Theodore de Bary, ed., *Living Legacies at Columbia* (New York: Columbia University Press), 151–164, 165–183.

17. Trustee Minutes, November 2, 1891.

18. Trustee Minutes, January 23, 1893.

19. On Columbia in early 20th century, McCaughey, *Stand, Columbia*, chaps. 7 – 11.

20. William Bradford, *Of Plymouth Plantation*, 1620–1647, ed. Samuel Eliot Morison (New York: Modern Library, 1952), 334.

21. On Chandler, see Robert L. Larson, "Charles Frederick Chandler: His Life and Work" (Columbia PhD, 1950): and Marston T. Bogert, "Charles Frederick Chandler, 1836–1925," *Biographical Memoir* (Washington, D.C.: National Academy of Sciences, 1931), 125–181.

22. Kurland, *Seth Low*.

23. On Chandler's favoring his students and assistants, Larson, "Chandler."

24. Larson, "Chandler,"

25. Larson, "Chandler."

26. Ibid.

27. Larson, "Chandler."

28. Seth Low to Charles Frederick Chandler, January 8 , 1890; Chandler to Low, January 23, 1890; Low to Chandler, November 24, 1891, Chandler Papers, Historical Subject File, CUA.

29. Low to Chandler, March 17, 1892, Chandler Papers; Larson, "Chandler," 340.

30. Chandler to Seth Low, February 2, 1891, Chandler Papers, Columbiana, vol. 11 (1890–1894).

31. Arthur Hixson, "The Chemical Engineering Department," a memorandum provided James Kip Finch in preparation of his 1954 history of the Engineering school article; Finch Papers, Columbia University Archives.

32. "Report of the Finance Committee," May 7, 1894, Box 3, School of Mines Records, Columbia University Archives.

33. Charles Frederick Chandler to Seth Low, January 28, 1895, Box 3, School of Mines Records; John Pine to Seth Low, February 7, 1895, John Pine Folder, Historical Subject File, CUA.

34. Low to Chandler, November 6, 1896, Chandler Papers, CUA.

35. Chandler to Low, January 4, 1897, Chandler Papers, CUA.

36. Seth Low to John Pine, February 8, 1895.

37. Michael Rosenthal, *Nicholas Miraculous: The Amazing Career of the Redoubtable Dr. Nicholas Murray Butler* (New York: Farrar, Straus and Giroux, 2006), chap. 5.

38. Larson, Chandler, 364.

39. On Chandler's retirement party, *Columbia University Quarterly* 12 (1910): 84-page supplement; Seth Low on list of guests but not in attendance, Box 8, Folder 11, School of Mines Records, CUA.

40. Leonard W. Fine and Ronald Breslow, "Chemistry at Columbia," ed. de Bary, *Living Legacies*, 199–209.

41. On Parsons crack about naptha launches, Nicholas Murray Butler to Frederick R. Hutton, February 26, 1903; on Hutton's extracted retirement, Butler to Hutton, November 2, 1906, Hutton Papers, Box 328, Central Files, CUA, Butler Library.

42. Frederick A. Goetz to Butler, December 19, 1907, Goetz Papers, CUA.

43. Goetz to John B. Pine, April 30, 1908, Goetz Papers.

44. Goetz to John B. Pine, March 22, 1907; William Barclay Parsons to Pine, April 2, 1907.

45. Butler to Goetz, January 30, 1908, April 21, 1908.
46. Goetz to John B. Pine, April 30, 1908, Goetz Papers, CUA.
47. Goetz to Frederick Keppel, January 13, 1909.
48. Goetz to John B. Pine, April 11, 1916, Goetz Papers, CUA.
49. Nicholas Murray Butler to Joseph W. Barker, December 17, 1931.
50. Charles R. Mann, *A Study of Engineering Education* (New York: Carnegie Foundation for the Advancement of Teaching, 1918), 4; reported on Harvard and Missouri trying 6-year program but then abandoning it.
51. Butler identified engineering alumni as the source of the idea on several occasions, including his "Remarks [Confidential] made at the meeting of the Trustees of Columbia University," May 3, 1926, 6.
52. Robert Stevens, "Two Cheers for 1870: The American Law School," *Perspectives in American History* 5 (1971), 405–548; Jerold S. Auerbach, *Unequal Justice: Lawyers and Social Change in Modern America* (New York: Oxford University Press, 1976).
53. Seth Low, *President's Annual Report*, October 7, 1895. Dartmouth's Thayer School of Engineering had a 6-year program.
54. Frederick Keppel to Butler, November 1, 1910, Keppel Papers, Columbiana. On Goetz's objections to merging admissions, August 25, 1909, Keppel Folder, Central Files Box 459, CUA. On Keppel scotching Goetz's idea for an undergraduate school of applied science, Keppel to Butler, November 1, 1910.
55. Butler to Goetz, January 7, 1910, Goetz Papers, Columbiana.
56. Butler to Goetz, November 7, 1910, Goetz Folder, Central Files, CUA.
57. Goetz to Butler, January 15, 1914.
58. Butler to Goetz, December 29, 1911; on Parsons, Goetz to Butler, April 19, 1913, Goetz Papers, Columbiana.
59. Frederick A. Goetz to Utah correspondent, May 1, 1914, Goetz Papers, Box 461, CUA.
60. Goetz to Utah alumnus, reported in Goetz to Butler, September 9, 1914, Frederick A. Goetz Folder, Box 401, Central Files, CUA.
61. Goetz to Butler, January 15, 1914, Box 461, CUA.
62. James Kip Finch, *Trends in Engineering Education: The Columbia Experience* (New York: Columbia University Press, 1948).
63. Ibid.
64. On academic anti-Semitism, see Harold S. Wechsler, *A History of Selective College Admission in America, 1870–1970* (New York: Wiley Interscience, 1977).
65. On the "Hebrew problem," Keppel to Butler, July 19, 1910, Box 459, Central File, CUA.
66. Wechsler, *Selective College Admission*; and McCaughey, *Stand, Columbia*, chap. 9.
67. Frederick Keppel to Nicholas Murray Butler, April 26, 1911, Keppel papers.
68. McCaughey, *Stand, Columbia*, 262–267; Frederick Keppel to Engineering School COI, April 26, 1911, October 26, 1911, Keppel Papers.
69. Keppel to Butler, May 13, 1913, Keppel Papers, Box 459, Central Files, CUA.
70. Butler to Keppel, January 27, 1914, Keppel Papers, Columbiana.

71. McCaughey, *Stand, Columbia*, 260–261.

72. Benjamin B. Lawrence, October 18, 1917, Benjamin Lawrence Folder, Box 329, Central Files, CUA.

73. Benjamin B. Lawrence to Frederick Keppel, December 19, 1909; Keppel to Lawrence, December 24, 1909, Lawrence Papers, Columbiana.

74. On Hawkes, McCaughey, *Stand, Columbia*, 267; Butler to George B. Pegram, June 29, 1922, Pegram Papers, Box 334, Central Files, CUA.

75. Lee Anna Embrey, "George Braxton Pegram, 1876–1958," *Biographical Memoir* (Washington, D.C.: National Academy of Sciences, 1970); on Pegram's reluctance to become dean, George B. Pegram to Butler, May 2, 1917, Pegram Papers, Box 334, Central Files, CUA.

76. Butler to Pegram, April 26, 1918; Pegram to Butler, May 3, 1918, Pegram Papers, Central Files, CUA.

77. Pegram to Butler, September 27, 1917, October 6, 1917, Pegram Papers, Central Files, CUA.

78. Butler to Pegram, June 29, 1922.

79. On Butler's openness to the idea of a woman being admitted to the engineering school, Butler to Pegram, November 19, 1917; on Pegram's opposition and that of "a considerable majority" of his faculty, Pegram to Butler, November 23, 1917.

80. Butler to Pegram, June 29, 1922.

81. On Finch-induced reforms, James Kip Finch, *Trends in Engineering Education: The Columbia Experience* 9 (New York: Columbia University Press, 1948).

82. H.G. Bowen, "History of the Navy Post-Graduate Engineering Course at Columbia University," in Charles E. Lucke, "Mechanical Engineering in Columbia University," January 1, 1950, copy in SEAS Mechanical Engineering Department.

83. [Milton Cornwell] *Report of the Alumni Committee on the Engineering Schools*, printed October 20, 1927. Included among the School of Mines Records, CUA.

84. Ibid.

85. Ibid.

86. [Confidential] Nicholas Murray Butler, "Remarks made at the meeting of the Trustees of Columbia University, May 3, 1926," Columbia University Archives, 1–4.

87. Ibid., 6–7.

88. Butler to Pegram, June 4, 1925, Pegram Papers, Box 334, CUA.

89. Gano Dunn to Butler, January 15, 1929; Dunn to William Barclay Parsons, July 16, 1929, Gano Dunn Folder, Box 354, Central Files, CUA.

90. Robert S. Woodward to Butler, January 19, 1904, School of Mines Records, CUA; F. E. Wright, "Robert Simpson Woodward, 1849–1924," *Biographical Memoirs* (Washington, D.C.: National Academy of Sciences, 1927).

91. Arthur W. Hixson, "The Department of Chemical Engineering," unpublished manuscript written in the early 1950s for the upcoming bicentennial history of Columbia University, Dwight Miner Papers, CUA.

92. Hixson, "Department of Chemical Engineering."

93. Nicholas Murray Butler to Frederick A. Goetz, February 12, 1915; Hixson, "Department of Chemical Engineering."

4. THE GREAT DEPRESSION AND THE GOOD WAR 1930–1945

1. Thomas H. Read to Joseph W. Barker, October 21, 1932, Barker Papers, Box 359, CUA.
2. Barker to Nicholas Murray Butler, October 30, 1935, Barker Papers.
3. Gospel of Luke, 10: 38–42, *New American Bible, Revised New Testament* (Grand Rapids: Eerdmans, 1988).
4. Paul Forman, "The Primacy of Science in Modernity, of Technology in Postmodernity, and of Ideology in the History of Technology," *History and Technology* 23 no. 1 and 2 (2007): 1–152.
5. Rudyard Kipling, "The Sons of Martha," *Rudyard Kipling: Selected Poetry* (London: Penguin, 1992).
6. John Dewey, "Needed—A New Politics," as reprinted in *The Later Works* 11:274–281, on 279; quoted in Forman, "Primacy of Science," 26.
7. Gano Dunn, "Speech to the Columbia Engineering Alumna Association," 1930, as quoted in Forman, "Primacy of Science," 14.
8. Nicholas Murray Butler, ["Confidential"] "Remarks to Trustees," May 3, 1926, Butler Papers.
9. Butler to Barker, December 23, 1931, Barker Papers, Box 359, CUA.
10. Gano Dunn to William Barclay Parsons, July 16, 1929, Gano Dunn Folder, Central Files, CUA.
11. Butler to Barker, January 13, 1932, Barker Papers, Box 359, CUA.
12. Morton Keller and Phyllis Keller, *Making Harvard Modern: The Rise of America's University* (New York: Oxford University Press, 2001), Part 1; Christopher Lécuyer, "The Making of a Science-Based Technological University: Karl Compton, James Killian, and the Reform of MIT, 1930–1957," *Historical Studies in the Physical and Biological Sciences* 23 (1992), 153–180.
13. Morris Bishop, *A History of Cornell* (Ithaca, NY: Cornell University Press, 1962), 476.
14. Michael Rosenthal, *Nicholas Miraculous: The Amazing Career of the Redoubtable Dr. Nicholas Murray Butler* (Farrar, Straus and Giroux, 2006), 434–460; McCaughey, *Stand, Columbia*, 320–327.
15. On Bush's role at MIT and Terman's role at Stanford, see Stuart W. Leslie, *The Cold War and American Science: The Military-Industrial-Academic Complex at MIT and Stanford* (New York: Columbia University Press, 1993); and Rebecca Lowen, *Creating the Cold War University: The Transformation of Stanford* (Berkeley: University of California Press, 1997).
16. Joseph W. Barker to Nicholas Murray Butler, May 20, 1931, Barker Papers, Box 359, CUA.
17. Memoir of James Kip Finch, "*Transactions of the American Society of Civil Engineers* (1967), 687.
18. Barker to Butler, October 30, 1935, Barker Papers.
19. Barker to Butler, May 23, 1934, Barker Papers.
20. James Kip Finch, "Research and Patent Policies," *Trends in Engineering Education: The Columbia Experience* (New York: Columbia University Press, 1948), 86–98.
21. Charles E. Lucke to Barker, January 31, 1931, Barker Papers.
22. Barker to Lucke, April 14, 1931; Lucke to Barker, April 18, 1931.
23. Butler to Barker, April 27, 1931; Barker to Butler, October 30, 1935, Barker Papers.
24. Barker to Butler, October 30, 1935, Barker Papers.

25. Yannis Tsividis, "Edwin Howard Armstrong: Pioneer of the Airwaves," ed. William Theodore De Bary, *Living Legacies at Columbia* (New York: Columbia University Press, 2006), 297–306. Armstrong's anticipated objections may have been a factor in Stanford's Frederick E. Terman, who expressed interest in coming to Columbia in 1947, not doing so. See Frank D. Fackenthal to James Kip Finch, July 20, 1946, Finch Papers.

26. Thomas H. Read to Barker, October 21, 1932, Barker Papers.

27. Barker to Butler, May 23, 1933, Barker Papers.

28. Ibid.

29. Edwin T. Layton Jr. *The Revolt of the Engineers: Social Responsibility and the American Engineering Profession* (Baltimore: Johns Hopkins University Press, 1986), 227–228.

30. James Kip Finch on Rautenstrauch's retirement in 1943. On Rautenstrauch as a teacher, Ira Millstein, interview with author, August 22, 2011.

31. Rautenstrauch's two interwar colleagues were less controversial. William S. Ayars had joined Rautenstrauch in the move from mechanical engineering in 1925 and stayed until 1939. Robert T. Livingston, an authority on public power, joined two years after its founding; he became head of Long Island Lighting Company, even as he stayed on at Columbia until retiring in 1965.

32. Frank Fackenthal to James Kip Finch, December 27, 1941, Finch Papers.

33. On Columbia in the 1930s, McCaughey, *Stand, Columbia*, chap. 11.

34. For the situation nationwide in the 1930s, Bruce Seeley, "Research, Engineering, and Science in American Engineering Colleges, 1900–1960," *Technology and Culture* 34 (April 1993): 344–386.

35. Butler to Barker, October 18, 1932; Barker to Butler, November 2, 1932, Barker Papers.

36. On students in the Depression, Robert Cohen, *When the Old Left Was Young: Student Radicals and America's First Mass Student Movement, 1929–1941* (New York: Oxford University Press, 1993).

37. On student life at Columbia College and the engineering school in the 1930s, Bernard Queneau, interview with author, June 15, 2011; Immanuel Lichtenstein, interview with author, August 25, 2011.

38. Bernard R. Queneau, BS 1932; MS 1933) , "Memoirs," unpublished manuscript provided by the author.

39. Hyman Rickover, "Remarks," the Egleston Arad Dinner, 1955.

40. Bernard R.Queneau, interview with author, June 2, 2012; Immanuel Lichetnstein, interviews with author, August 25, 2011, May 31, 2013.

41. On Columbia's 75th Anniversary, James Kip Finch, ed., *"The School of Engineering, 1864–1939,"* Columbia University Bulletin of Information, June 1, 1940.

42. Randall Jarrell, "In Those Days." *Complete Poems* (New York: Farrar, Straus & Giroux, 1969).

43. McCaughey, *Stand, Columbia*, 425; Rosenthal, *Nicholas Miraculous*, 435–438.

44. Fon W. Boardman, *Columbia: An American University in Peace and War* (New York: Columbia University Press, 1944).

45. Ibid.

46. Co-ed Stanford enrolled three times as many women as men during World War II.

47. James Kip Finch memo, September 8, 1942; Finch to Frank D. Fackenthal, January 5, 1943.

48. Gloria Reinish, interview with author, June 4, 2011. Her son is James Reinish (BS 1982), the daughter Nancy Passow (BS chemical engineering 1972), the granddaughter, Ariel Reinish (BS bioengineering, 2010).

49. Boardman, *Columbia: An American University in Peace and War*, James Kip Finch, "The School of Engineering in Wartime," September 21, 1943, Finch Papers.

50. Boardman.

51. H. G. Bowen, "History of the Navy Post-Graduate Engineering Course at Columbia," in Charles Edward Lucke, "Mechanical Engineering in Columbia University" (unpublished manuscript, 1950), copy in SEAS Department of Mechanical Engineering.

5. MISSING THE BOAT 1945–1964

1. Frank D. Fackenthal to James Kip Finch, February 7, 1945, Finch Papers.

2. Finch to Fackenthal, January 28, 1948, Finch Papers.

3. Stuart W. Leslie, *The Cold War and American Science: The Military-Industrial-Academic Complex at MIT and Stanford* (New York: Columbia University Press, 1993); Robert A. McCaughey, *International Studies and Academic Enterprise: A Chapter in the Enclosure of American Learning* (New York: Columbia University Press, 1984), 113–166.

4. McCaughey, *Stand, Columbia*, chap. 12.

5. Barker to Fackenthal, March 29, 1946, Barker Papers.

6. McCaughey, *Stand, Columbia*, 328.

7. Travis B. Jacobs, *Eisenhower at Columbia* (New Brunswick: Transaction Publishers, 2001), 43–44.

8. Ibid.; McCaughey, *Stand, Columbia*, 338–341.

9. Daniel Beshers, interview with author, April 17, 2012.

10. James Kip Finch to Frank Fackenthal, January 1, 1948, , Finch Papers, CUA.

11. Story recounted in William J. McGill, "Reminiscences," Oral History Collection, CUA.

12. Finch to Fackenthal, January 28, 1948, Finch Papers, CUA.

13. Finch to Fackenthal, October 16, November 7, 1946, Finch Papers, CUA.

14. Albert C. Jacobs to Finch, March 25, 1949, Finch Papers, CUA.

15. Finch to Jacobs, August 1, 1949, Finch Papers, CUA.

16. On Dunning's appointment, Jacobs, *Eisenhower at Columbia*, 228.

17. Herbert L. Anderson, "John Ray Dunning, 1907–1975," *Biographical Memoir* (Washington, D.C.: National Academy of Sciences, 1989), 162–186; Morton Friedman, interview with author, July 13, 2011.

18. Richard Rhodes, *The Making of the Atomic Bomb* (New York: Simon & Schuster, 1986), 268–270, 297–298, 332–333; Samuel Devons, "I. I. Rabi: Physics and Science at Columbia, in America, and Worldwide," William Theodore de Bary, *Living Legacies at Columbia* (New York: Columbia University Press, 2006), 211–236.

19. On Rabi's possible role, Robert A. Gross, interview with author, March 26, 2013.

20. Daniel Beshers, interview with author, April 17, 2012.

21. William J. McGill, "Reminiscences," recorded in 1980 by Henry Graff, 92.

22. Morton Friedman, interview with author, July 13, 2011.

23. On AEC and universities, Roger L. Geiger, *Research and Relevant Knowledge: American Research Universities Since World War II* (New York: Oxford University Press, 1993), 20–23; W. W. Havens Jr., Report of the Committee on the Educational Program in Nuclear Engineering, April 11, 1974.

24. Folder on "TRIGA Reactor," Box 702, Central File, CUA; contains a 1960–1974 chronology and related documents.

25. On Brookhaven, Daniel J. Kevles, *The Physicists: The History of a Scientific Community in Modern America* (Cambridge: Harvard University Press, 1995), 368–74; on Nevis, Devons, "Rabi," ed. de Bary, *Living Legacies*, 229.

26. TRIGA Reactor chronology, Box 702, Central Files, CUA.

27. "100 attend 3-hour reactor debate," *Columbia Spectator*, April 29, 1977.

28. Report of the Division of Nuclear Science and Engineering, 1974.

29. "Hennessy as Executive Dean," *Columbia Spectator*, September 28, 1964.

30. On Dunning, author's interview with Seymour Melman, November 1997.

31. "Henry Krumb," Mining Hall of Fame Inductee Database, Leadville, Colorado: http://www.mininghalloffame.org's rise.

32. Ibid.

33. Paul Duby, interview with author, June 23, 2011.

34. Grayson Kirk, "Oral History," 8, p. 1080, Oral History Collection, Columbia University.

35. Text of Krumb will provided author by Professor Ponisseril Somasundaran, June 26, 2012.

36. Ibid.

37. Kirk, "Oral History," February 13, 1959. The deceased Mudd had no connection with Columbia, though his two sons, Harvey Seeley Mudd (EM, 1912) and Dr. Seeley Greanleaf Mudd (BS 1919) were graduates of the engineering school.

38. $500,000 during Krumb's lifetime; $6.3 million upon his death in 1958; $8.2 million on his wife's death in 1962.

39. Edward Leonard, interview with author, July 14, 2011; on Harris's second thoughts, Paul Duby, interview with author, February 2, 2012; Zvi Galil, interview with author, June 12, 2012, and subsequent correspondence; Ponisseril Somasundaran, interview with author, June 26, 2012.

40. Paul Duby, interview with author, February 2, 2012, June 12, 2013.

41. Polykarp Kusch to William J. McGill, July 1970.

42. http://engineering.columbia.edu/leverlongenough

43. Ibid.

44. Ibid.

45. Ages when not available were calculated based on birth occurring 21 years before receipt of baccalaureate degree.

46. Birthplace of faculty when not known assumed to be in region where undergraduate education occurred.

47. Gerard Ateshian, interview with author, August 23, 2012.

48. National Academy of Engineering was founded in 1964 under the same congressional act that led to the founding of the National Academy of Sciences.

49. Case for "golden age" made in author interviews with Christian Meyer, July 13, 2011; René Testa, August 24, 2011; and Frank DiMaggio, August 24, 2011.

50. Bernard Roth, "Editorial: In Memory of Professor Ferdinand Freudenstein," *Journal of Mechanical Design* 129 (March 2007): 241–242. Contains an autobiographical sketch written in 2000.

51. On Lotfi Zadeh, http://www.eecs.berkeley.edu/Faculty/Homepages/zadeh.html.

52. Damian Doria, "Interview with Dr. Cyril Harris," September 12, 2007, Niels Bohr Library and Archive, http://www.aip.org/history/ohilist/31674.html.

53. On Drew, "50 Chemical Engineers of the 'Foundation Age," *CEP Magazine* (September 2008).

54. Anne E. Bromley, "In Memoriam: Elmer Gaden," *UVA Today*, March 15, 2012.

55. On Kellogg, "Mr. Herbert H. Kellogg," *NAE Members Directory*, http://www.nae.edu/29098.aspx; "Tributes to Nathaniel Arbiter, 1911–2008," *Minerals Engineering International*, October 7, 2008.

56. Herbert Deresiewicz, "Raymond D. Mindlin—A Bio/Bibliographical Sketch," http://www.olemiss.edu/sciencenet/mindlin/mindlin-bio.pdf. Mindlin may have been, upon his appointment as assistant professor in 1940, the first Jew to be a member of the engineering school faculty.

57. Paul Goldberger, "Mario G. Salvadori, Engineer and Inner-City Teacher, 90," *The New York Times*, June 28, 1997.

58. Mario G. Salvadori, "Hans Heinrich Bleich, 1909–1985," *Memorial Tributes: National Academy of Engineering* 3 (1989): 45–49.

59. "Bruno A. Boley: Biography," *Journal of Thermal Stresses* 15 (1992): v.

60. Harold Liebowitz, "Alfred Martin Freudenthal, 1906–1977," *Memorial Tributes: National Academy of Engineering* 1 (1977): 63–65.

61. Shu Chien, "Richard Skalak, 1923–1997," *Memorial Tributes: National Academy of Engineering* 9 (2001): 254–259.

62. "David B. Hertz," Wikipedia, http://en.wikipedia.org/wiki/David_B._Hertz; John Ralph Ragazzini, obituary, *The New York Times*, November 24, 1988; "John Ralph Ragazzini," The Mathematics Genealogy Project: http://genealogy.math.ndsu.nodak.edu/id.php?id=95228.

63. On the preponderance of émigré civil engineers, Jeremy Isenberg, "Melvin L. Baron," *Memorial Tributes: National Academy of Engineering* 10 (2002): 15.

64. Edward Leonard, interview with author, July 14, 2011; Morton Friedman, interview with author, May 19, 2011.

65. Jacques Barzun, *The American University* (New York: Harper & Row, 1968), 34–40.

66. John R. Thelin, *A History of American Higher Education* (Baltimore: Johns Hopkins University Press, 2004), 260–316.

67. John S. Rigden, *Rabi: Scientist and Citizen* (New York: Basic Books, 1987), 73.

68. Morton Friedman, interview with author, May 19, 2011.

69. On Columbia's intellectual climate in the postwar period, Carolyn G. Heilbrun, *When Men Were the Only Models We Had: My Teachers Barzun, Fadiman, Trilling* (Philadelphia: University of Pennsylvania Press, 2002).

70. Rigden, *Rabi*, 73.

71. Devons, "Rabi," in ed. de Bary, *Living Legacies*.

72. Lionel Trilling, "The Leavis-Snow Controversy," in *Beyond Culture: Essays on Literature and Learning* (New York: Harcourt Brace Jovanovich, 1979), 126–154; Quentin Anderson, "Lionel Trilling at Columbia," ed. William Theodore de Bary, *Living Legacies at Columbia* (New York: Columbia University Press, 2006), 3–12.

73. Jacques Barzun, *The House of Intellect* (New York: Harper & Brothers, 1959), 19–21; Barzun, *Science; The Glorious Entertainment* (New York: Harper & Row, 1964), 4–8, 6; Thomas Vinciguerra, "Jacques Barzun: Cultural Historian, Cheerful Pessimist, Columbia Avatar," ed. de Bary, *Living Legacies*, 383–395.

74. Wallace S. Broecker, Interview with the Lamont-Doherty Project, May 8, 1977, Oral History Collection, Butler Library, Columbia University.

75. Barzun, *Science: The Glorious Entertainment*; Daniel Bell, *The Coming of Post-Industrial Society* (New York: Basic Books, 1973), 26.

76. Trilling, "Leavis-Snow," 126–154, 143.

77. Ibid., 139; Friedman, interview with author, May 19, 2011.

78. Stephen S. Visher, *Scientists Starred, 1903–1943, in American Men of Science* (Baltimore: Johns Hopkins University Press, 1947); David S. Webster, "A Note on Very Early Academic Quality Ranking by James McKeen Cattell," *Journal of the History of Behavioral Sciences* 20 (April 1984): 180–183.

79. Raymond M. Hughes, *Report of the Committee on Graduate Education* (Washington, D.C.: American Council on Education, 1934); Hayward Keniston, *Graduate Study and Research in the Arts and Science at the University of Pennsylvania* (Philadelphia: University of Pennsylvania Press, 1959); Allan Cartter, *An Assessment of Quality in Graduate Education* (Washington, D.C.: American Council on Education, 1966).

80. Cartter, *Assessment of Quality*, 100–107.

81. Polykarp Kusch, "Report on Problems of the University," to William J. McGill, July 8, 1970, Polykarp Kusch Papers, Central Files, CUA.

82. On "quality of faculty" as a lagging indicator, Kenneth D. Roose and Charles J. Andersen, *A Rating of Graduate Programs* (Washington, D.C.: American Council on Education, 1970), 9.

83. Cartter, *Assessment of Quality*.

84. Peter Blau and Rebecca Margulies, "America's Leading Professional Schools," *Change* 5 (November 1973): 21–27; Blau and Margulies, "The Reputations of American Professional Schools," *Change* 6 (Winter 1974–1975): 42–47.

85. Frank da Cruz, Columbia University Computing History, http://www.columbia.edu/cu/computinghistory/; see also Bruce Gilchrist, "Computing at Columbia University: A Brief Historical Review," unpublished, December 1979.

86. "Herman Hollerith," IBM Archives, http://www-03.ibm.com/ibm/history/exhibits/builders/builders_hollerith.html; Hollerith's first tabulating machine is described in *The School of Mines Quarterly*.

87. Frank da Cruz, Columbia Computing Timeline, http://www.columbia.edu/cu/computinghistory/#timeline.

88. Jean Ford Brennan, *The IBM Watson Laboratory at Columbia University: A History* (IBM, 1971).

89. Frank da Cruz, Columbia Computing Timeline.

90. Ibid.; Herbert R. J. Grosch, *Computer: Bit Slices from a Life* (Novato, CA: Third Millennium Books, 1991).

91. Ibid.

92. F. S. Beckman, Computer Science Activity at Columbia University, October 20, 1967, Charles Babbage Institute Archives; William Aspray, "Was Early Entry a Competitive Advantage? US Universities That Entered Computing in the 1940s," *IEEE Annals of the History of Computing* (July–September 2000), 42–87, 66–72.

93. Mary Bellis, "Timeline of IBM History," http://inventors.about.com/od/timelines/tp/Timeline-IBM-History.htm.

94. Edward Leonard, interview with author, July 14, 2011; Joseph F. Traub, interview with author, January 30, 2011; William Aspray, "An Interview with Joseph F. Traub," Charles Babbage Institute, Center for the History of Information Processing, University of Minnesota, April 5, 1984, 21–25.

95. Harold S. Wechsler, *The Qualified Student: A History of Selective College Admission in America, 1870–1970* (New York: Wiley, 1977). Graduates of New York City engineering programs who later taught at Columbia include John Ragazzini, Frederick Winter, Dudley Fuller, Herbert Goldstein, Bruno Boley, Walter La Pierre, Seymour Melman, Eugene Machlin, Jerome Weiner, Herbert Deresiewicz, Morton Friedman, Robert Bernstein, Stephen Unger, Herbert Schorr, James Galligan, Mischa Schwartz, Joseph Traub.

96. Fritz Stern, interview with author, December 19, 2002.

97. "Frosh Engineering Students Participate in College Program," *Columbia Spectator*, September 17, 1959.

98. "Dean Fears Engineers Strain College Facilities," *Columbia Spectator*, February 23, 1960.

99. "Drop-outs Plague Engineers," *Columbia Spectator*, May 13, 1963.

100. Tullio (Ted) Borri, interview with author, June 3, 2011.

101. On women students, see http://engineering.columbia.edu/leverlongenough. On NROTC and Regents Scholarships, author direct experience with both.

102. Gloria Brooks Reinish, interview with author, June 4, 2011; Dr. Wallace Broecker, interview with Spenser Weart, Lamont-Doherty Earth Observatory, NY, November 14, 1997.

103. "A Farewell to Camp Columbia," *Engineering News* (Fall 2001).

104. Camp Columbia and the launching of *Pulse*. Tullio (Ted) Borri, interview with author, May 31, 2011.

105. Steven Bellovin, interview with author, May 8, 2012.

106. Andrew S. Dolkart, *Morningside Heights: A History of Its Architecture and Development* (New York: Columbia University Press, 1998), 199–200.

6. BOTTOMING OUT 1965-1975

1. Polykarp Kusch to William McGill, July 8, 1970, Folder 16, Central Files.
2. Editorial, "Enough Is Enough," *Pulse*, September 27, 1973.
3. On views of Gordon Brown, dean emeritus of MIT, Walter Rosenblith to President McGill, October 4, 1973, Central Files.
4. McCaughey, *Stand, Columbia*, chap. 5.
5. On McGill's triaging strategy and his presidency, William J. McGill, "Reminiscences," interviews with Henry Graff in 1979–1980, Oral History Collection, Butler Library, Columbia University.
6. Grayson Kirk, "Oral History," interviews with Chauncey Olinger, 1985–1988, Oral History Collection, Butler Library, Columbia University.
7. David Horowitz, *Radical Son: A Generational Odyssey* (New York: Free Press, 1997); James Kunen, *The Strawberry Statement: Notes of a College Revolutionary* (New York: Random House, 1969); McCaughey, *Stand, Columbia*, 423–461.
8. Morton Friedman, interview with author, May 19, 2011.
9. Benjamin B. Lawrence to William Barclay Parsons, November 6, 1917, Box 329, Historical Subjects Files.
10. Hennessy, "The Quiet Revolutionaries" Class Day, 1965, School of Mines Papers, CUA.
11. On military ties, see Chapter 4.
12. SEAS as an unlikely source of student revolutionaries, Peter Likins, *A New American Family: A Love Story* (Tucson: University of Arizona Press, 2011), 80–81.
13. On Seymour Melman, *Pulse*, March 14, 1968; on John Englund, *Pulse*, November 16, 1967; on Victor Paschkis, *The New York Times*, June 20, 1991; on Mario G. Salvadori, *Columbia University Record* 23, no. 2 (September 12, 1997).
14. On engineers and their views on student protests on campus, Bureau of Applied Social Research, "Student and Faculty Response to the Columbia Crisis: A Preliminary Report Based on Partial Returns," May 20, 1968, mimeograph, Columbia University Archives; on the disruptions, Jerry Avorn and Robert Friedman, *Up Against the Ivy Wall: A History of the Columbia Crisis* (New York: Atheneum, 1969); McCaughey, *Stand, Columbia*, 442–461.
15. Preliminary arrest statistics provided in Robert Brookhart to Grayson Kirk, May 4, 1968, Columbia Crisis Papers, CUA; McCaughey, *Stand, Columbia*, 456–459.
16. Allen H. Barton, "The Columbia Crisis: Campus, Vietnam, and the Ghetto," *Public Opinion Quarterly* 32 (Fall 1968): 333–351; "[SEAS] Faculty First to Act on Strike," *Pulse*, May 6, 1968.
17. On MIT, Stuart W. Leslie, "'Time of Troubles' for the Special Laboratories," in ed. David Kaiser, *Becoming MIT: Moments of Decision* (Cambridge: MIT Press, 2010), 123–144; on Cornell, Cushing Stout, *Divided We Stand; Reflections on the Crisis at Cornell* (New York: Doubleday, 1971); on Wisconsin and the campus scene generally, Willis Rudy, *The Campus and a Nation in Crisis* (Madison, NJ: Fairleigh Dickinson University Press, 1996), 192.

18. Among the casualties, the School of Library Science, the linguistics and the geography departments, *Columbia University Quarterly*.

19. On enrollment drops and dropouts, *Pulse*, May 1, 1968; Leonard DiFioree to Wesley Hennessy, November 14, 1971, Faculty Engineering Minutes, 1964–1971.

20. Robert Brookhart to William T. de Bary, November 19, 1974, Box 208, Office of the Provost Records.

21. Kenneth D. Roose and Charles J. Andersen, *A Rating of Graduate Programs* (Washington, D.C.: American Council on Education, 1970).

22. Ibid., 36–98.

23. Ibid., 100–107; Polykarp Kusch, "Memorandum re SEAS," December 1970, Box 25, Historical Subjects File, CUA.

24. Peter Blau and Rebecca Margulies, "America's Leading Professional Schools," *Change* 5 (November 1973): 21–27; Blau and Margulies, "The Reputations of American Professional Schools," *Change* 6 (Winter 1974–1975): 42–47.

25. *Columbia Spectator*, September 8, 1969.

26. *Columbia Spectator*, April 8, 1968.

27. McCaughey, *Stand, Columbia*, 497–498.

28. McGill, "Reminiscences."

29. William J. McGill, *The Year of the Monkey: Revolt on Campus, 1968–69* (New York: McGraw-Hill, 1982).

30. McGill, "Reminiscences," June 26, 1979.

31. Polykarp Kusch, "Report on Problems of the University," July 8, 1970, Office of the Provost Files, CUA.

32. Ibid.

33. Polykarp Kusch to Wesley Hennessy, June 30, 1971, Polykarp Kusch Folder, Office of the Provost Files, CUA.

34. McGill, "Reminiscences," 88.

35. Ibid.

36. As reported at the SEAS faculty meeting of April 25, 1973, Central Files (1973).

37. Ibid., 499.

38. William T. de Bary to Wesley Hennessy, Office of the Provost Files, CUA.

39. De Bary to Hennessy, January 10, 1973, Box 712, Central Files, CUA.

40. Wesley Hennessy to James Young, February 24, 1972, Office of the Provost Files; *Pulse*, February 23, 1972.

41. On dispute over the teaching of differential equations, *Spectator*, October 26, 1972; on Columbia College calls for merger, *Spectator*, December 2, 1970; "Enough Is Enough," *Pulse*, September 27, 1973.

42. Wesley Hennessy to William J. McGill, January 18, 1974, Box 732, Central Files.

43. Rosalind Rosenberg, *Changing the Subject: How the Women of Columbia Shaped the Way We Think About Sex and Politics* (New York: Columbia University Press, 2004), 245–257.

44. Ibid., 256.

45. "Engineering's First Woman Faculty Member," *Pulse*, December 17, 1970.

46. "[President] Ford to City: Drop Dead," [New York] *Daily News*, October 30, 1975, front page; Vincent J. Cannato, *The Ungovernable City: John Lindsay and His Struggle to Save New York* (New York: Basic Books, 2002).
47. William T. de Bary to Wesley Hennessy, Office of the Provost Records.
48. Elmer Gaden to William T. de Bary, March 2, 1973, Central Files; Committee Chair Ralph Schwartz defended the second report in letter to President McGill, July 8, 1973, Central Files.
49. Application numbers in April 19, 1973, Admissions Report, Central Files; Minutes of SEAS Faculty meeting, April 25, 1973, Central Files; Daniel Beshers, interview with the author, April 17, 2012.
50. William J. McGill to William O. Baker, July 17, 1973, Central Files.
51. William J. McGill, memorandum on de Bary response to "SEAS Aims and Goals," October 4, 1973, Central Files.
52. McGill to Walter Rosenblith, September 6, 1974, Central Files; Hennessy informed the SEAS faculty of his intention to retire in early 1975 on September 16, 1974, Box 753, Central Files; Hennessy on SEAS faculty wanting his job, in McGill memo of July 23, 1973.

7. CATCHING A LIFT 1976–1980

1. Peter Likins to SEAS Faculty, October 11, 1976, Box 791 (1976–1977), Central Files.
2. William Aspray, "Interview with Joseph F. Traub," March 29, 1985, Charles Babbage Institute, Center for the History of Information Processing, University of Minnesota, 22.
3. "McGill Denies Report on Engineering Dean Search," *Columbia Spectator*, July 25, 1975; William T. de Bary, memo, March 3, 1976, Box 776, Central Files.
4. On Likins coming to search committee's attention, Richard Longman, interview with author, November 2, 2011.
5. Peter Likins, interview with author, January 19, 2012.
6. Likins, interview, January 29, 2012; Peter Likins, *A New American Family: A Love Story* (Tucson: University of Arizona Press, 2011), 1–59.
7. President McGill to SEAS Faculty, April 13, 1976, Box 772, Central Files.
8. Robert Brookhart to Provost William T. de Bary, February 12, 1974, Office of Provost Records.
9. Peter Likins to William T. de Bary, August 16, 1976, Office of Provost Records.
10. Peter Likins to SEAS Faculty, October 16, 1976; McGill, "Reminiscences," 66.
11. Likins on Columbia Engineering Affiliates, Box 828, Central Files; Likins, *A New American Family*, 80–81.
12. Peter Likins to Provost Michael Sovern, March 14, 1979, Box 828, Central Files.
13. Peter Likins to Bruce Bassett, Box 809 (1978–1979), Central Files; Peter Likins to William T. de Bary, March 1, 1978, Office of the Provost Records.
14. Peter Likins to William T. de Bary, "Conspectus of the School of Engineering," December 7, 1977, Box 385, Office of the Provost Records.
15. Mischa Schwartz, interview with author, May 9, 2012; Donald Goldfarb, interview with author, June 20, 2012; Ponisseril Somasundaran, interview with author, June 26, 2012;

Nickolas Themelis, interview with author, June 5, 2012; Richard Longman, interview with author, November 11, 2011.

16. McCaughey, *Stand, Columbia*, 498–501.

17. Gerald Navratil, interview with author, June 25, 2012; Christian Meyer, interview with author, July 13, 2011.

18. Joseph F. Traub, interview with author, January 30, 2011; Jonathan Gross, interview with author, June 11, 2012.

19. Peter Likins, interview with author, January 25, 2012; on opposition from electrical engineering, Mischa Schwartz, interview with the author, May 9, 2012; for Rabi remark about "science," Traub interview, January 30, 2011.

20. Traub interview, January 30, 2011.

21. Traub interview, January 30, 2011; Likins interview, January 25, 2012.

22. Traub interview, January 30, 2011; Jonathan Gross, interview with author.

23. David Yao, interview with author, June 27, 2011.

24. Peter Likins, interview with author, January 25, 2012. The faculty member was likely Herbert Goldstein.

25. Gerald Navratil, interview with the author, June 25, 2012.

26. Likins, *A New American Family*, 42.

27. Likins, interview with author, January 29, 2012.

28. William Aspray, "Interview with Joseph F. Traub," March 29, 1985, Charles Babbage Institute, The Center for the History of Information Processing, University of Minnesota, 22.

29. William J. McGill, "Reminiscences," Oral History Collection, Henry Graff interviewer, Rare Books and Manuscripts Library, Columbia University, 1980.

30. McGill and academic freedom, "Bankers Association Speech," April 19, 1977, Box 820 ["TRIGA"], Central Files; McGill to faculty opposing activation, January 26, 1976, Box 782, Central Files, CUA.

31. McGill, "Reminiscences" 2, 185; B. Drummond Ayres, "Three Mile Island: Notes from a Nightmare," *The New York Times*, April 16, 1979, 1A.

32. William McGill, "Reminiscences" 2, 185.

33. Ibid., 186.

34. McCaughey, *Stand, Columbia*, 549; Likins, interview, January 25, 2012.

35. Peter Likins, interview with author, January 29, 2012.

8. UNEVEN ASCENT 1980–1994

1. Michael I. Sovern to Ralph J. Schwarz, November 5, 1981, Box 865, Central Files.

2. Christopher Durning, interview with author, November 1, 2012.

3. McCaughey, *Stand, Columbia*, 534–535.

4. Michael I. Sovern, *An Improbable Life: My Sixty Years at Columbia and Other Adventures* (New York: Columbia University Press, 2014).

5. Jonathan Gross, interview with author, March 26, 2013; Zvi Galil, interview with author, June 12, 2012.

6. Salvatore Stolfo, interview with author, May 27, 2012.

7. Joseph F. Traub, interview with author, January 30, 2011; Joseph F. Traub, interview with William Aspray, March 29, 1985, Charles Babbage Institute, University of Minnesota, 16.

8. Traub, interview with Aspray, 24.

9. Traub, interview with Aspray; Jonathan Gross, interview with author, June 11, 2012; Salvatore Stolfo, interview with author, May 27, 2012.

10. Zvi Galil, interview with author, June 12, 2012.

11. Ibid.

12. Ibid.

13. Jonathan Gross, interview with author, June 11, 2012.

14. William Aspray, "Was Early Entry a Competitive Advantage? US Universities That Entered Computing in the 1940s," *IEEE Annals of the History of Computing* (July-September, 2000): 42–87.

15. Charles E. Lucke, "Mechanical Engineering in Columbia University," bound typescript, January 1, 1950, provided author by Richard Longman, Department of Mechanical Engineering, SEAS.

16. On MIT's "Tech Plan," see Christophe Lécuyer, "Patrons and a Plan," in ed. David Kaiser, *Becoming MIT: Moments of Decision* (Cambridge: MIT Press, 2010), 59–80.

17. See chap. 2, on Newberry's problems with extra-mural employment, p. 40.

18. Charles H. Townes, "A Life in Physics," 1991–1992 interviews conducted by Suzanne B. Riess, Regional Oral History Office, Bancroft Library, University of California, Berkeley, California (1994), 166–168.

19. Daniel M. Dumych, "Pupin, Michael I," *American National Biography Online*, http://www.anb.org/articles/13/13–01339.html?a=1&n=Pupin%2C%20Michael&ia=-at&ib=-bib&d=10&ss=0&q=1; Frederick L. Rhodes, *Beginnings of Telephony* (1929).

20. On Armstrong's inventions, Yannis Tsividis, "Edwin Howard Armstrong: Pioneer of the Airwaves," in William T. de Bary, ed., *Living Legacies at Columbia* (New York: Columbia University Press, 2005), 297–306; Charles H. Townes, "A Life in Physics," 196–197.

21. On Bayh-Dole Act of 1980, Jonathan R. Cole, *The Great American University* (New York: Public Affairs, 2009), 163–171.

22. Ibid., 168.

23. Not all licensing arrangements between faculty and the university worked to everyone's satisfaction. Salvatore Stolfo, interview with author, May 27, 2012.

24. On the entrepreneurial turn among research universities in the 1980s, Roger L. Geiger, *Research and Relevant Knowledge: American Research Universities Since World War II* (New York: Oxford University Press, 1993), 316–326.
   According to Orin Herskowitz, executive director and VP, Columbia Technology Ventures, as of 2013 Columbia generates over 300 research-based inventions, leading to 200 patents, 70+ licenses, and 15+ start-ups each year. Correspondence with author, August 7, 2013.

25. Gerald Navratil, interview with author, June 25, 2012; Michael Mauel, interview with author, May 23, 2012; Robert A. Gross, interview with author, March 26, 2013.

26. Michael I. Sovern to Ralph J. Schwarz, November 5, 1981, Box 865, Central Files.

27. Grayson Kirk to faculty members considering forming the Institute of Bioengineering at Columbia, March 23, 1967, Box 749, Central Files; Edward Leonard, interview with author, July 14, 2011.

28. William J. McGill to William L. Nastuk, July 14, 1971, Box 749, Central Files.

29. Van C. Mow, interview with author, August 8, 2011; Shu Chien, "A Tribute to Professor Van C. Mow: A Wonderful Scholar and Leader in Bioengineering," *Cellular and Molecular Bioengineering* 2, no. 3 (September 2009): 282–284.

30. Mow interview, August 28, 2011.

31. Ibid.; Gerard Ateshian, interview with author, August 23, 2012.

32. Mischa Schwartz, interview with author, May 9, 2012.

33. Schwartz interview, May 9, 2012; "The Center for Telecommunications Research at Columbia University," *CTR Newsline* 4, no. 2 (Fall 1992).

34. Irving P. Herman, correspondence and interview with author, August 16, August 20, 2013.

35. Robert A. Gross, interview with author, March 26, 2013; Gerald Navratil, interview with author, June 25, 2012.

36. Gross interview, March 26, 2013.

37. Ibid.

38. Zvi Galil, interview with author, June 12, 2012.

39. Rosalind Rosenberg, *Changing the Subject, How the Women of Columbia Shaped the Way We Think About Sex and Politics* (New York: Columbia University Press, 2004).

40. Gloria Reinish (née Brooks), interview with author, June 4, 2011; Elena Robbins, interview with author, August 24, 2011; Anna K. Longobardo, interview with author, May 31, 2013.

41. For women graduates of SEAS, 1953, see http://engineering.columbia.edu/leverlongenough.

42. Elena Robbins, interview with author, August 24, 2011; Edward Leonard, interview with author, July 14, 2011.

43. Kathleen McKeown, interview with author, September 25, 2012; Gail Kaiser, interview with author, August 28, 2012; Julia Hirschberg, interview with author, September 6, 2012.

44. On Neumark and Chan, Paul Duby, interview with author, February 2, 2012.

45. David H. Auston, interview with author, April 2, 2013.

46. Ibid.

47. Sovern, *An Improbable Life*, 220–222; McCaughey, *Stand, Columbia*, 552–556.

48. Auston interview, April 2, 2013.

49. The small size of SEAS departments was a recurrent concern of SEAS accreditation committees. See Engineers' Council for Professional Development to William McGill, November 14, 1971, SEAS Files; Accreditation Board for Engineering and Technology, Inc., to Michael I. Sovern, August 21, 1992, SEAS Files.

50. For Auston's consolidation plan, Carl Gryte, "Chemical Engineering at Columbia University: Celebrating 100 Years, 1905–2005," CD-ROM, 2005; "SEAS Dean Proposes Closing a Department," Columbia *Spectator*, February 17, 1992.

51. Christopher Durning, interview with author, November 1, 2012; Paul Duby, interview with author, February 2, 2012; Ponisseril Somasundaran, interview with author, June 26, 2012.

52. Nicholas Themelis, interview with author, June 5, 2012; Christopher Durning, interview with author, November 1, 2012.

53. Gryte, "Chemical Engineering"; "SEAS Dean Proposes Merging a Department," *Columbia Spectator*, May 13, 1992.

54. Columbia *Spectator*, April 11, 1991; Auston interview, April 2, 2013.

55. Morton Friedman, interview with author, May 19, 2011; Paul Duby, interview with author, February 2, 2012, Ponisseril Somasundaran, interview with author, June 26, 2012.

56. Van C. Mow, interview with author, August 8, 2011; Josph Ienuso, interview with author, March 12, 2013.

57. Thus, Columbia College's applicant to enrolled ratio during this period was about 12 to 1; SEAS's 4.4 to 1. See Appendix 2, SEAS Undergraduates Admissions Data. http://engineering .columbia.edu/leverlongenough

58. B. Lindsay Lowell, et al., "Steady as She Goes? Three Generations of Students Through the Science and Engineering Pipeline," October 2009, Institute for the Study of International Migration [ISIM], Georgetown University, http://policy.rutgers.edu/faculty/salzman/ SteadyAsSheGoes.pdf.

59. On College skepticism over the prospect of merged admissions, Eric Furda, interview with author, February 18, 2013.

60. Ibid.; Joseph Ienuso, interview with author, March 12, 2013.

61. Jan Krukowski Associates, "Columbia University—Columbia College/School of Engineering & Applied Science, Student Recruitment Marketing Plan," draft submitted October 18, 1991; Jan Krukowski, correspondence with author, May 8, 2013.

62. Ibid., 12–13.

63. Ibid., 19.

64. Ibid., 28, 48

65. Ibid., 44, 66.

66. Furda interview, February 18, 2013; for recent SEAS Undergraduate Admissions Data, see Digital Appendix, SEAS website.

67. Joseph Ienuso, interview with author, March 12, 2013.

68. Auston interview, April 2, 2013; Van C. Mow, interview with author, August 8, 2011; Christopher J. Durning, interview with author, November 1, 2012.

9. A SCHOOL IN FULL 1995–2007

1. Nicholas Turro, interview with author, February 29, 2012.

2. *Engineering News* (Spring 2007).

3. Gerald Navratil, interview with author, June 25, 2012.

4. Lauren Goodman, "Galil Named SEAS Dean after Year Long Search," *Columbia Spectator*, July 19, 1996.

5. Morton Friedman, "School's Constituencies Reflect on Galil Deanship," *Engineering News*, Spring 2007, 1.

6. Galil, interview with author, June 12, 2012.

7. Ibid. Income from the school's online instructional program, which by the late 1990s amounted to $2 million annually, also helped the school's finances.

8. For faculty growth, see http://engineering.columbia.edu/leverlongenough; on start-up funding, Christopher Durning, interview with author, November 1, 2012. Galil was notoriously parsimonious when it came to furnishing the dean's office, retaining a couch that had seen service since Mudd opened in 1962.

9. Salvatore Stolfo, interview with author, May 27, 2012.

10. This historical summary based on author interviews with Steve Bellovin, May 8, 2012; Zvi Galil, June 12, 2012; Jonathan Gross, June 11, 2012; Julia Hirschberg, September 6, 2012; Gail Kaiser, August 28, 2011; Kathleen McKeown, September 28, 2012; Shree Nayar, June 26, 2012; Salvatore Stolfo, May 27, 2012; Joseph Traub, January 30, 2011; Mihalis Yannakakis, December 13, 2011.

11. Steven Bellovin, interview with author, May 8, 2012; Julia Hirschberg, interview with author, September 6, 2012; Zvi Galil, interview with author, June 12, 2012; Mihalis Yannakakis, interview with author, December 13, 2012.

12. On hiring practices, author interview with Salvatore Stolfo, May 27, 2012; Sanat Kumar, interview with author, June 12, 2013.

13. On gender implications of undergraduate preferences, Kathleen McKeown, interview with author, September 25, 2012; Julia Hirschberg, interview with author, September 6, 2012; Gail Kaiser, interview with author, August 28, 2012.

14. Ponisseril Somasundaran, interview with author, June 26, 2012.

15. Christopher Durning, interview with author, November 1, 2012; Alan West, interview with author, March 4, 2013.

16. Nicholas Turro, interview with author, February 29, 2012.

17. Academy members included Herbert Kellogg, Nathaniel Arbiter, Ponisseril Somasundaran, Nickolas Themelis.

18. Nickolas Themelis, interview with author, June 5, 2012; Paul Duby, interview with author, February 2, 2012.

19. Paul Duby, interview with author, June 12, 2013.

20. Duby interview, February 2, 2012.

21. Peter Schlosser, interview with author, June 17, 2013.

22. Duby interview, June 12, 2013.

23. Robert A. Gross, interview with author, March 26, 2013; Gerald Navratil, interview with author, June 25, 2012.

24. John Chu, interview with author, April 16, 2013. Fu also funded the Fu Professorship of Applied Mathematics, which went to Chu.

25. Ibid.

26. Michael Mauel, interview with author, May 23, 2012.

27. The Stormer/Pinczuk arrangements also included their claim on the top floor of Schapiro Hall for laboratory space.

28. Mauel interview, May 23, 2012.

29. Jonathan R. Cole, "Strategic Planning Commission: Reports of the Task Forces," October 1993 McCaughey, *Stand, Columbia*, 557–561.

30. Van C. Mow, interview with author, August 8, 2011; Gerard Ateshian, interview with author, August 23, 2012.

31. Mow interview, August 8, 2011.

32. Ibid.

33. X. Edward Guo, interview with author, March 28, 2013.

34. Ibid.

35. David Yao, interview with author, June 27, 2012; Donald Goldfarb, interview with author, June 20, 2012.

36. Goldfarb interview, June 20, 2012.

37. Yao interview, June 27, 2012.

38. Gallego interview, April 8, 2013.

39. Yao interview.

40. Ibid.; Gallego interview; Emanuel Derman, *My Life as a Quant: Reflections on Physics and Finance* (New York: Wiley, 2004).

41. Gallego interview; Anne Marie O'Neill, interview with author, April 23, 2013.

42. Gallego interview; Garud Iyengar, interview with author, September 19, 2013.

43. Walter A. Lapierre, "Department of Electrical Engineering," typescript written in early 1950s in preparation for the Bicentennial History of Columbia University, Dwight Miner Papers, CUA.

44. Charles Zukowski, interview with author, May 31, 2013; Mischa Schwartz, interview with author, May 9, 2012.

45. Zvi Galil, interview with author, June 12, 2012.

46. Zukowski interview; correspondence with author, September 22, 2013.

47. My knowledge of the recent history of electrical engineering supplemented by correspondence with Dimitris Anastassiou and Keren Bergman.

48. Recent history of mechanical engineering has been provided in interviews with Richard Longman, November 11, 2011; Gerard A. Ateshian, August 23, 2013; and Jeffrey W. Kysar, June 18, 2013; and correspondence with Michael Lai. Recent history on civil engineering based on interviews with Morton Friedman, May 19, 2011; Frank DeMaggio, August 24, 2011, and December 12, 2012; Rene Testa, August 24, 2011; Raimondo Betti, December 17, 2012; and George Deodatis, August 20, 2013.

49. On Crow, author interview with Frank DiMaggio, August 24, 2011. Michael Crow is presently president of Arizona State University.

50. Michael Lai, correspondence with author, April 2013; Richard Longman, interview with author, November 11, 2011; Raimondo Betti, interview with author, December 17, 2012.

51. Gerard Ateshian, interview with author, August 23, 2012.

52. Raimondi Betti, interview with author, December 17, 2012; George Deodatis, interview with author, August 19, 2013.

53. George Deodatis, correspondence with author, August 12, 2013.

54. On Rupp and undergraduate push, McCaughey, *Stand, Columbia,* 573–577.

55. Eric Furda, interview with author, February 18, 2013; Joseph Ienuso, interview with author, March 12, 2013; "Record High Applications: Good News Only for the Best," *Engineering News,* Fall 2001.

56. On Galil's on-campus visibility, Morton B. Friedman, "School's Constituencies Reflect on Galil Deanship," *Engineering News,* Spring 2007.

57. Zvi Galil, interview with author, June 12, 2012.

58 Ibid.

59. "Dean Galil Becomes President of Tel Aviv U," *Engineering News,* Spring 2007.

60. Alan Leshner, AAAS Address, February 17, 2002, quoted in Paul Forman, "The Primacy of Science in Modernity, of Technology in Postmodernity, and of Ideology in the History of Technology," *History and Technology* 23 (April 2007), 11.

61. Forman, "Primacy of Science," 2.

62. Paul Forman, "Into Quantum Mechanics: The Maser as 'Gadget' of Cold-War America," in Paul Forman and J. M. Sanchez-Ron, eds., *National Military Establishments and the Advancement of Science and Technology* (Boston: Kluwer Academic, 1996), 261–326.

63. On Rabi's changing his mind, Mischa Schwartz, interview with author, May 9, 2012.

64. William J. McGill, Oral History interview, June 26, 1979.

65. Nonscience provosts include William H. Carpenter (1912–1927); Frank Diehl Fackenthal (1937–1946); Albert C. Jacobs (1947–1949); Grayson L. Kirk (1949–1953); John A. Krout (1953–1958); Jacques Barzun (1958–1967); David B. Truman (1967–1969); Peter B. Kenen (1969–1970); William Theodore de Bary (1971–1978); Michael I. Sovern (1979–1980); Jonathan R. Cole (1989–2003); Alan Brinkley (2003–2009).

66. On Goldberger's interest in the work of engineers, Salvatore Stolfo, interview with author, May 27, 2012.

67. Jonathan R. Cole, *The Great American University* (New York: Public Affairs Books, 2009), 91–101.

68. Morton Friedman, "School's Constituencies Reflect on Galil Deanship," *Engineering News,* Spring 2007.

69. On fundamental science losing its previous agency in the 1990s, Daniel J. Kevles, "Preface, 1995: The Death of the Superconducting Super Collider in the Life of American Physics," in Daniel J. Kevles, *The Physicists: The History of a Scientific Community in Modern America* (Cambridge: Harvard University Press, 1995), ix–xlii.

70. 38 of the 170 engineering faculty (22%) profiled in *Excellentia Eminentia Effectio* (New York: Fu Foundation School of Engineering and Applied Science, 2011) were classified as engaged in research related to health.

71. Thomas Cech, 2004, as quoted in Forman, "Primacy of Science," 11; *Rosalind Williams, Retooling: A Historian Confronts Technological Change* (Cambridge: MIT Press, 2002), xi.

72. Eugene Galanter, "The Cultures of Academe," Conference on The Sciences at Columbia, 2003, 1–2.

10. A LEVER LONG ENOUGH: SEAS AT ONE HUNDRED FIFTY

1. Archimedes and the lever, as quoted on the wall in the lobby of Mudd, 5th floor.
2. Salvatore Stolfo, interview with author, May 27, 2012.
3. Anna K. Longobardo, interview and correspondence with author, May 30, 2013, September 30, 2013.
4. http://engineering.columbia.edu/leverlongenough.
5. The absence of research activity was not peculiar to Columbia's interwar engineering faculty, but the norm among engineering faculties before WWII. See Bruce Seeley, "Research, Engineering, and Science in American Engineering Colleges, 1900–1960," *Technology and Culture* 34, no. 2 (April 1993): 344–386, 366.
6. http://engineering.columbia.edu/leverlongenough.
7. Morton Friedman, interview with author, May 19, 2011.
8. Of the 139 faculty hired between 2001 and 2013, 27 (19%) have since departed.
9. For postwar applied scientists, see chap. 5.
10. There is a nice irony in the fact that these departments were brought into being 35 years ago by Dean Peter Likins, who insisted upon his coming to Columbia three years earlier that "I was a real engineer." Hardly anyone hired since and still here would today so identify herself. It was the outgoing Likins who hired Richard Osgood Jr., an electrical engineer at MIT's Lincoln Lab and very much of the Gross research-oriented persuasion, to fill the faculty slot in electrical engineering and one in applied physics being vacated by Gross becoming dean. Osgood's arrival in 1981 provided further evidence that a fundamental and likely irreversible change was under way within the SEAS faculty.
11. Roger L. Geiger, *Research and Relevant Knowledge: American Research Universities Since World War II* (New York: Oxford University Press, 1993); Hugh Davis Graham and Nancy Diamond, *The Rise of American Research Universities* (Baltimore: Johns Hopkins University Press, 1997); Jonathan R. Cole, *The Great American University* (New York: Public Affairs, 2009).
12. Based on author interviews with 41 faculty and five deans. See "Interviews Conducted by Author," Digital SEAS History Appendix.
13. "Best Engineering Schools in 2013," *U.S. News and World Report*—http://grad-schools. usnews.rankingsandreviews.com/best-graduate-schools/top-engineering-schools/eng-rankings
14. Shree Nayar, Rocco Servedio, Michael Collins, "Columbia SEAS Ranking: 2001–2011," PowerPoint Presentation, January 27, 2012; Shree Nayar, interview with author, June 26, 2012.
15. Julia Hirschberg, interview with author, September 6, 2012.
16. Count as of September 2013.
17. Natural language processing group includes Kathleen McKeown, Vision/computer graphics group includes Shree Nayar, Cybersecurity group includes Steve Bellovin, Gail Kaiser Tal Malkin, Angelos Keromytis, Sal Stolfo; on making the top-10 "very achievable," author interview with Shree Nayar, June 26, 2013; on Yannakakis, http://scholar.google.com/

citations?user=_pPy-pAAAAAJ&hl=en&oi=ao; on Institute for Data Sciences & Engineering, see *The Columbia University Record* 38, no. 09 (April 2013): 1.

18. *Quant News*, *Google Scholar*; on Mechanical Engineering Department, author interviews with Richard Longman, November 11, 2011; Gerard Ateshian, August 3, 2012; Jeffrey Kysar, June 18, 2013.

19. On traditional standing of physical sciences at Columbia, Arthur W. MacMahon, *Future of the University* (New York: Columbia University Press, 1957).

20. The six science departments: biological sciences, chemistry, earth and environmental science (then geology), ecology, evolution and environmental biology, mathematics, and physics.

21. Columbia's cognate science departments in 2013 have been credited with 21 science faculty members as Nobel laureates, plus another 11 graduates. This does not include Nobelists in physiology and medicine (23), economics (11), literature (3), or peace (2).

22. The count as of June 2013.

23. Information on sponsored research at Columbia from Carol Tycko, director of sponsored research, April 10, 2013, correspondence, June 28, 2013; James Kemp; Fred Kant, interview with author, July 16, 2013; correspondence with Michael Purdy, executive vice president for research; Deborah Stiles, vice president for research, interview with author, April 8, 2013.

24. SEAS Expenditures by Fiscal Year, as reported to *USNWR*, provided by William Yandolino, September 19, 2013.

25. http://engineering.columbia.edu/leverlongenough.

26. Department comparisons are not precise, due to uncertainty about number of faculty budgeted in each department, but this is not likely to affect the rank ordering or the comparisons made with cognate science departments.

27. Jeffrey Kysar, interview with author, June 18, 2013.

28. Salvatore Stolfo, interview with author, May 27, 2012; Kysar interview.

29. Robert Zussman, *Mechanics of the Middle Class: Work and Politics Among American Engineers* (Berkeley: University of California Press, 1985); Peter Likins, *A New American Family: A Love Story* (Tucson: University of Arizona Press, 2011),

30. Author interviews with David Yao, June 27, 2011; Salvatore Stolfo, May 27, 2012; Karl Sigman, June 4, 2013; Fred Kant, July 16, 2013.

31. Victoria Hamilton, interview with author, April 10, 2013.

32. "Entrepreneurship Alumni Mentors," *Columbia Engineering News*—http://engineering.columbia.edu/entrepreneurship-alumni-mentors.

33. *Columbia Engineering Bulletin*, 2011–2012, 193; Frank DiMaggio, interview with author, August 24, 2011.

34. E. M. Forster, *Howards End* (London: Putnam's, 1911).

35. Shree Nayar, interview with author, June 26, 2012; Karl Sigman, interview with author, June 4, 2013.

36. Shree Nayar, interview with author.

37. Most joint appointments are limited to senior faculty because junior appointees are seen as best represented by a single department in the tenuring process. Twenty-three senior

members have appointments in two departments. All nine departments have at least one member who holds a joint appointment; APAM has five members.

38. Five SEAS departments (APAM, CompSci, EE, EEE, IEOR) have members who are also members of a Columbia science department (APAM and EEE have two each).

39. Eleven SEAS faculty hold appointments in one of Columbia's other professional schools. Seven of these are appointments of members of the biomedical engineering department with departments in the health sciences.

40. On Earth Institute, Peter Schlosser, interview with author, June 17, 2013; on CMSRS, Venkat Venkatasubramanian, correspondence with author, July 2013.

41. On regional labs, Charles Zukowski, interview with author, May 31, 2013; George Deodatis, interview with author, August, 21, 2013.

42. SEAS faculty engaged in neural engineering include Barclay Morrison III, Dimitris Anastassiou, Elizabeth M. C. Hillman, Aural A. Lazar, Paul Sajda.

43. McCaughey, *Stand, Columbia*, 392–394, 499–500.

44. The Earth Institute of Columbia University—http://www.earth.columbia.edu/articles/view/1791.

45. Lenfest Center for Sustainable Energy created in 2004 with a $15 million gift from university trustee Gerry Lenfest (Law 1958)—http://energy.columbia.edu/; Paul Duby, interview with author, June 12, 2013.

46. Gerard Ateshian, interview with author, August 23, 2012.

47. Chester Lee, interview with author, May 31, 2013.

48. Digital SEAS History Appendix—Students.

49. "Statistical Abstract," Office of the Provost, Planning and Institutional Research, Columbia University—http://www.columbia.edu/cu/opir/abstract.html.

50. Ibid.

51. On student occupational choices, Cliff Massey, "Columbia Entrpreneurs Talk New Business in NYC," December 20, 2013, http://engineering.columbia.edu/columbia-entrepreneurs-talk-new-business-nyc.

52. Steven Bellovin (SEAS 1972), interview with author, May 8, 2012; Christopher Durning (SEAS BS 1978), interview with author, September 1, 2012.

53. Joke provided by professor of computer science Salvatore Stolfo's website: The graduate with a science degree asks, "Why does it work?" The graduate with an engineering degree asks, "How does it work?" The graduate with an accounting degree asks, "How much will it cost?" The graduate with an arts degree asks, "Do you want fries with that?"

54. SEAS faculty in 2013 as SEAS undergraduates: Paul Diament (1963); Steven Bellovin (1972); Christopher Durning (1978); Siu-Wai Chan (1980); David Vallancourt (1981); Michael Hill (1983); a Columbia College undergraduate, Chris H. Wiggins (BA 1993).

55. F. Augustus Schermerhorn (EM 1870) 1877–1908; Lenox Smith (EM 1868) 1883–1913; William Barclay Parsons (CE 1882) 1897–1932; Benjamin B. Lawrence (EM 1878) 1909–1921; William Fellows Morgan (EM 1884) 1910–1943; Arthur S. Dwight (EM 1885) 1916–1946; Frederick Coykendall (CE 1897) 1916–1956; Walter H. Aldridge (EM 1887)

1922–1959; Henry Hobart Porter (EM 1886) 1924–1947; Gano Dunn (EE 1891) 1928–1934; Milton L. Cornell (CE 1905) 1930–1936; Francis Blossom (CE 1891) 1935–1941 ; Edmund A. Prentis (EM 1906), 1938–44; Henry Krumb (EM 1898), 1941–47; Harris K. Masters (EM 1894), 1944–1950; Walter H. Sammis (EE 1917) 1947–1953; Robert D. Lilley (BS 1935), 1968–1980; Anna K. Longobardo (BS 1949, MS 1952), 1990–1996; Armen A. Avanessians (MS 1983), 1998–2010; Savio Tung (BS 1973), 1998–2011; Vikram Pandit (BS 1976, MS 1977), 2003. Changes in board rules implemented after 1968 have since limited the tenure to a maximum of two 6-year terms. Twelve are chosen by the trustees acting alone, six by the trustees working with the executive committee of the University Senate, and six elected, one each year, after nomination by the university's alumni.

56. Among SEAS's biggest benefactors: Z. Y. Fu, Chinese businessman, brother-in-law of Professor John Chu–1997—$26 million Henry Krumb (EM 1898)–1958, 1962—$16 million Percy K. Hudson (EM 1899)–1977—$12 million

57. Anna K. Longobardo, interview with author, May 31, 2013; Daniel K. Libby, interview with author, May 31, 2013. On the Barnard pre-engineering program, Virginia C. Gildersleeve, *Many a Good Crusade* (New York: Macmillan, 1954), 104.

58. On McGill's appreciation of Lilley's counsel, McGill, "Reminiscences," Oral History Collection, CU; on Lilley's service to SEAS, see Peter Likins, *A New American Family: A Love Story* (Tucson; University of Arizona Press, 2011), 55.

59. On Egleston medalists, http://www.seas.columbia.edu/alumni/egmeds.html.

60. Columbia faculty as Egleston medalists include Edwin H. Armstrong (1939), Jewell M. Garrelts (1972), Donald Burmister (1978), Elmer L. Gaden Jr. (1986), Masanobu Shinozuka (2004), Guy Longobardo (2006), Lotfi Zadeh (2007), http://cuengineeringalumni.org/ceaa-awards/egleston-medal/.

61. For Samuel Johnson medalists, see http://cuengineeringalumni.org/ceaa-awards/samuel-johnson-medal/(A third, the Pupin Medal, honors scientists irrespective of their academic origins.)

62. Peggy Maher, associate dean of advancement and alumni affairs, correspondence with author, August 28, 2013; Daniel Libby, ex-president of Columbia Engineering Alumni Association, interview with author, May 31, 2013; "Back to School: Alumni Reconnect at Reunion," *Engineering News,* June 19, 2013.

63. Jonathan Gross, interview with author, June 11, 2012.

64. For SEAS deans, see Digital SEAS History—Administration.

65. See chap. 7 on McGill restructuring.

66. Galil's deanship discussed in chap. 9.

67. On the accomplishments of the Gross deanship, Gerald Navratil, interview with author, June 25, 2012.

68. Lisa Foderaro, "An Immigrant's Journey to a Top Post at Columbia," *The New York Times*, June 18, 2009.

69. Ibid.

70. Andrew Gaspar, interview with author, May 31, 2013.

71. Foderaro, "An Immigrant's Journey."

72. Editorial Board, "Grading the administrators: Dean Feniosky Peña-Mora," *Columbia Spectator*, November 9, 2010.

73. Richard Pérez-Peña, "Discord over Dean Rocks Columbia Engineering School," *The New York Times*, December 7, 2011.

74. Ibid.

75. https://bwog.com/tag/pena-mora/.

76. Ariel Kaminer, "After Revolt, a Dean at Columbia Steps Down," *The New York Times*, July 3, 2012.

77. Lillian Chen, "MIT Prof Boyce Named SEAS Dean," *Columbia Spectator*, March 27, 2013; on MIT dean's remark, Walter Rosenblith to William McGill, October 4, 1973.

78. Letter to Coatsworth quoted in Pérez-Peña, "Discord," *The New York Times*, December 7, 2011. One unbidden response for the 145-year-old School of Engineering and Applied Science of Columbia University in the city of New York might well have been the counter question: "What am I, chopped liver?"

79. Patrick McGeehan, "By Deadline, 7 Bids in Science School Contest," *The New York Times*, October 31, 2011.

80. Richard Pérez-Peña, "Two Top Suitors Are Emerging for New Graduate School of Engineering," *The New York Times*, October 16, 2011; Staff Editorial, "SEAS the Day," *Columbia Spectator*, October 24, 2011; Finn Vigeland, "CU Plays Up City Ties in Bid for Funds," *Columbia Spectator*, October 28, 2011, 1.

81. Richard Pérez-Peña, "In Mayor's Contest for a New Engineering School, Only One Judge Counts," *The New York Times*, November 7, 2011; Richard Pérez-Peña, "After Yearlong Competition, Mayor Is Said to Pick Cornell for Science School," *The New York Times*, December 19, 2011.

82. Eric F. Newcomer, "Columbia Gets $15 Million to Expand a School," *The New York Times*, July 30, 2012.

83. "The Recent Difficulties in Columbia College," [Boston] *Christian Examiner*, July, 1854.

84. On New York City as "Silicon Alley," see http://en.wikipedia.org/wiki/Silicon_Alley.

85. Goldie Blumenstyk, "The New Technology Campus: Handicapping the Contenders," *The Chronicle of Higher Education*, October 23, 2011.

86. On the importance of these full-time off-ladder faculty, Julia Hirschberg, interview with author, September 6, 2012. They include Aaron Kyle, BME; Michael Hill, ChemE; José Sánchez, Civil; Adam Cannon, CompS; Robert Farrauto, EEE; David Vallancourt, EE; Martin Haugh, IEOR; Fred Stolfi, ME.

87. "Columbia Engineering Tops US News Report Best Online Graduate Engineering Program, http://engineering.columbia.edu/columbia-engineering-tops-us-news-best-online-grad-engineering-programs; Nicole Dyer, "Making Sense of MOOCs," *Columbia Engineering*, Spring 2013, 4–9, 9.

A BIBLIOGRAPHIC NOTE

A statistical appendix containing numerical data and lists is available at http://engineering
.columbia.edu/leverlongenough. This historical account of Columbia's engineering school has
relied primarily on three kinds of sources at three different points in the narrative. For the opening
chapters and occasionally thereafter I have drawn upon secondary accounts by historians of
American higher education (including myself), American engineering, and American technology.
These included:

Monte A. Calvert, *The Mechanical Engineer in America, 1830–1910: Professional Cultures in
     Conflict* (Baltimore: Johns Hopkins University Press, 1967);
Roger L. Geiger, *To Advance Knowledge: The Growth of American Research Universities* (New York:
     Oxford University Press, 1986);
——, *Research and Relevant Knowledge: American Research Universities Since World War II* (New
     York: Oxford University Press, 1993);
John F. Kasson, *Civilizing the Machine: Technology and Republican Values in America, 1776–1900*
     (New York: Penguin, 1976);
David F. Noble, *America by Design: Science, Technology, and the Rise of Corporate Capitalism*
     (New York: Oxford University Press, 1977).

I wish to acknowledge the early work of James Kip Finch, longtime professor and dean of the
Columbia engineering school, particularly for his accounts bearing directly on Columbia's engineers. These writings include:

Early Columbia Engineers: An Appreciation (New York: Columbia University Press, 1929);
"Engineering and Science: A Historical Review and Appraisal," *Technology and Culture* 2 (Fall
     1961): 318–332;

A History of the School of Engineering, Columbia University (New York: Columbia University Press, 1954);

Trends in Engineering Education: The Columbia Experience (New York: Columbia University Press, 1948).

Other secondary accounts relevant to the history of engineering and engineering education in America that helped me with the overall interpretive arc of my narrative include:

Paul Forman, "The Primacy of Science in Modernity, of Technology in Postmodernity, and of Ideology in the History of Technology," *History and Technology* 23, no. 1 and 2 (2007): 1–152;

Daniel J. Kevles, *The Physicists: The History of a Scientific Community in Modern America* (with new preface) (Cambridge: Harvard University Press, 1995);

Edwin Layton, "Mirror-Image Twins: The Communities of Science and Technology in 19th-Century America," *Technology and Culture* 12 (October 1971): 562–580;

Edwin T. Layton Jr., *The Revolt of the Engineers: Social Responsibility and the American Engineering Profession* (Baltimore: Johns Hopkins University Press, 1986).

Specific accounts of American engineering programs that I found useful in fashioning my own included:

David Kaiser and his colleagues writing on MIT in David Kaiser, ed., *Becoming MIT: Moments of Decision* (Cambridge: MIT Press, 2010);

Stuart W. Leslie, *The Cold War and American Science: The Military-Industrial-Academic Complex at MIT and Stanford* (New York: Columbia University Press, 1993);

Robert C. McMath Jr., *Engineering the New South: Georgia Tech, 1885–1985* (Athens: University of Georgia Press, 1985);

Rosalind Williams, *Retooling: A Historian Confronts Technological Change* (Cambridge: MIT Press, 2002).

I have also relied here on the work of several Columbia colleagues with whom over the years I have joined in a collaborative effort to illuminate our university's past:

Barry Bergdoll, *Mastering McKim's Plan: Columbia's First Century on Morningside Heights* (Miriam and Ira D. Wallach Art Gallery, Columbia University in the City of New York, 1997);

Jonathan R. Cole, *The Great American University* (New York: Public Affairs, 2009);

William Theodore de Bary, ed., *Living Legacies at Columbia* (New York: Columbia University Press, 2006);

Andrew S. Dolkart, *Morningside Heights: A History of Its Architecture & Development* (New York: Columbia University Press, 1998);

Robert A. McCaughey, *Stand, Columbia: A History of Columbia University in the City of New York, 1754–2004* (New York: Columbia University Press, 2003);

Rosalind Rosenberg, *Changing the Subject: How the Women of Columbia Shaped the Way We Think About Sex and Politics* (New York: Columbia University Press, 2004);

Michael Rosenthal, *Nicholas Miraculous: The Amazing Career of the Redoubtable Dr. Nicholas Murray Butler* (New York: Farrar, Straus and Giroux, 2007).

I wish also to note the importance for any interloper into the history of science and technology of the journal of the Society of the History of Technology, *Culture and Technology*, and the historical anthology edited by Terry S. Reynolds, *The Engineer in America* (Chicago: University of Chicago Press, 1991).

The middle chapters (2 to 6) of this account drew heavily upon the archival materials available in the Columbia University Archives, under the auspices of the Rare Book and Manuscript Library, Columbia University. Of particular use were the following:

Central Files;
Engineering Dean's Office Records;
Historical Biographical Files;
Historical Subjects Files;
Minutes of the Faculty of Engineering, 1907–1973;
Trustee Minutes.

In addition, I enjoyed access to the recently processed School of Mines Records and the Records of The Office of the Provost. Materials from all these archival files were made available through the late 1980s, when standing restrictions precluded further access. The Columbia University Archives has also been my repository of choice for *The School of Mines Quarterly*, *The Miner*, annual catalogs, and yearbooks of graduating classes of the Columbia Mines/Engineering school dating back to the 1880s.

The most recent chapters (7 to 10) relied primarily upon 96 interviews conducted with SEAS administrators, faculty, alumni, and students, as well as with Columbia administrators and faculty knowledgeable about SEAS. My practice in these interviews was to digitally record them, draw up a rough transcript of the interview, and submit it to the interviewee for corrections and elaboration. These subsequent exchanges often produced additional information of use. A full listing of the interviews by subject and date follows this note. I have also drawn selectively in these most recent chapters upon the serial publications of the School of Engineering and Applied Science, including *Columba Engineering Quarterly* and its successor, *Columbia Engineering*, and its recent *Excellentia, Eminentia, Effectio* (New York: Columbia Engineering, 2011).

FACULTY INTERVIEWS BY DEPARTMENT, 2009–2013

Irving	Hermann	Faculty/Current	APAM	20-Aug-13
Michael	Mauel	Faculty/Current	APAM	23-May-12
Gerald	Navratil	Faculty/Current	APAM	25-Jun-12
Richard	Osgood	Faculty/Current	APAM	18-Sep-12
C. K. John	Chu	Faculty/Retired	APAM	16-Apr-13
Edward	Guo	Faculty/Current	BME	28-Mar-13
Henry	Hess	Faculty/Current	BME	15-Mar-13
Andrew	Laine	Faculty/Current	BME	15-Mar-13
Van C.	Mow	Faculty/Current	BME	8-Aug-11
Van C.	Mow	Faculty/Current	BME	20-Dec-11
Van C.	Mow	Faculty/Current	BME	7-Mar-12
Christopher	Durning	Faculty/Current	ChemE	1-Nov-12
Sanat	Kumar	Faculty/Current	ChemE	12-Jun-13
Edward	Leonard	Faculty/Current	ChemE	14-Jul-11
Alan	West	Faculty/Current	ChemE	4-Mar-13
Nicholas	Turro	Faculty/Deceased	Chemistry	24-May-12
Raimondo	Betti	Faculty/Current	Civil	17-Dec-12
George	Deodatis	Faculty/Current	Civil	20-Aug-13
Morton B.	Friedman`	Faculty/Current	Civil	12-Nov-09
Mort	Friedman	Faculty/Current	Civil	19-May-11
Mort	Friedman	Faculty/Current	Civil	11-Jun-13
Christian	Meyer	Faculty/Current	Civil	13-Jul-11
Rene	Testa	Faculty/Current	Civil	24-Aug-11
Frank	DiMaggio	Faculty/Retired	Civil	24-Aug-11
Frank	DiMaggio	Faculty/Retired	Civil	12-Dec-12
Steve	Bellovin	Faculty/Current	CompSci	8-May-12
Jonathan	Gross	Faculty/Current	CompSci	11-Jun-12
Julia	Hirschberg	Faculty/Current	CompSci	6-Sep-12
Gail	Kaiser	Faculty/Current	CompSci	28-Aug-12
Kathy	McKeown	Faculty/Current	CompSci	25-Sep-12
Shree	Nayar	Faculty/Current	CompSci	26-Jun-12
Salvatore	Stolfo	Faculty/Current	CompSci	27-May-12
Joseph	Traub	Faculty/Current	CompSci	30-Jan-12
Joseph	Traub	Faculty/Current	CompSci	16-Feb-12
Mihalis	Yannakakis	Faculty/Current	CompSci	13-Dec-12
Robert A.	Gross	Faculty/Dean	Dean/ APAM	26-Mar-13

Zvi	Galil	Faculty/Dean	Dean/CompSci	12-Jun-12
David H.	Auston	Faculty/Dean	Dean/EE	2-Apr-13
Peter	Likins	Faculty/Dean	Dean/ME	24-Jan-12
Yannis	Tsividis	Faculty/Current	EE	22-Nov-11
Charles	Zukowski	Faculty/Current	EE	31-May-13
Mischa	Schwartz	Faculty/Retired	EE	10-Nov-09
Mischa	Schwartz	Faculty/Retired	EE	9-May-12
Paul	Duby	Faculty/Current	EEE	2-Feb-12
Paul	Duby	Faculty/Current	EEE	23-Jun-11
Paul	Duby	Faculty/Current	EEE	12-Jun-13
Peter	Schlosser	Faculty/Current	EEE	17-Jun-13
Ponisseril	Somasundaran	Faculty/Current	EEE	26-Jun-12
Daniel	Beshers	Faculty/Retired	EEE	17-Apr-12
Nickolas J.	Themelis	Faculty/Retired	EEE	5-Jun-12
Garud N.	Iyengar	Faculty/Current	IE/OR	17-Sep-13
Guillermo	Gallego	Faculty/Current	IEOR	8 Apr 13
Donald	Goldfarb	Faculty/Current	IEOR	10-Jul-13
Karl	Sigman	Faculty/Current	IEOR	4-Jun-13
David	Yao	Faculty/Current	IEOR	27-Jun-11
Gerard	Ateshian	Faculty/Current	ME/BME	23-Aug-12
Jeffrey	Kysar	Faculty/Current	ME	18-Jun-13
Michael	Lai	Faculty/Retired	ME	
Richard	Longman	Faculty/Current	ME	11-Nov-11

INTERVIEWS OF SEAS AND COLUMBIA ADMINISTRATORS

FIRST NAME	LAST NAME	POSITION	TENURE/ROLE	INTERVIEW
David H.	Auston	SEAS Dean	1990–1994	2-Apr-12
Mary C.	Boyce	SEAS Dean	2013–	20-Aug-13
Morton	Friedman	SEAS Vice Dean	1995–2012	12-Nov-09
Eric	Furda	CU Admin	Admissions 1990s	18-Feb-13
Zvi	Galil	SEAS Dean	1995–2007	12-Jun-12
Robert	Gross	SEAS Dean	1981–1990	26-Mar-13
Victoria	Hamilton	CU Admin	CU/RISE	10-Apr-13
Orin	Herskowitz	CU Admin	Columbia Technology Ventures*	6-Aug-13
Joseph	Ienuso	CU Admin	Admissions 1990s	12-Mar-13
Fred	Kant	CU Admin	Columbia Technology Ventures	16-Jul-13
Margaret	Kelly	SEAS Admin	Communications	9-Dec-09
James	Kemp	CU Admin	CU/Finance	17-Apr-13
Peter	Likins	SEAS Dean	1976–1980	19-Jan-12
Margaret	Maher	SEAS Admin	Alumni Affairs	7-Mar-13
Feniosky	Peña-Mora	SEAS Dean	2009–2012*	
Remi	Silverman	CU Admin	University Counsel*	19-Sep-13
Deborah	Stiles	CU Admin	VP Research	8-Apr-13
Carol	Tycko	CU Admin	Supported Research	10-Apr-13

*Correspondence.

SEAS ALUMNI INTERVIEWS

		1ST DEGREE	2ND DEGREE	3RD DEGREE	INTERVIEW DATE
Andrew	Gaspar	1969			31-May-11
Chester	Lee	19780			28-Mar-13
K. Daniel	Libby	1982	MS 1984		31-May-13
Immanuel	Lichtenstein	1943			25-Aug-11
Immanuel	Lichtenstein	1943			31-May-13
Anna K.	Longobardo	1949	MS 1952		31-May-13
Leonor	Lopez	1983			3-May-11
Ralf	Mekle		MS 1996	PhD 2003	31-May-13
Ira M.	Millstein	1947			22-Aug-11
Ben	Nineji	1982		PhD 1984	31-May-13
Nancy	Passow	1972			16-Aug-11
Bernard R.	Queneau	1933			2-Jun-12
Gloria	Reinish	1945	MS 1948	PhD 1974	4-Jun-11
Elena	Robbins	1948			24-Aug-11
Ingrid	Rodriquez	2001			6-Jul-11
Harvey	Rubin	1965	MS	DES 1976	3-May-11
Dan	Schiavello	1975	MS 1976		31-May-13
Alan M.	Silberstein	1969			31-May-13
Hitoshi	Tanaka	1963	MS 1976	DES 1976	31-May-13
Noah	Whitehead	2013			15-Mar-13

INDEX